SUPERSTUFF!

SUPERSTUFF!

MATERIALS THAT HAVE CHANGED OUR LIVES

FRED BORTZ

FRANKLIN WATTS
NEW YORK ■ LONDON ■ TORONTO ■ SYDNEY
A VENTURE BOOK ■ 1990

TO MY PARENTS,
WHO PRODUCED THE RAW MATERIAL,

TO MY TEACHERS,
WHO PROCESSED IT WITH CARE,

AND TO JEAN REYNOLDS,
WHOSE "FAN LETTER" AND ENCOURAGEMENT
PUT ME ON THE ROAD THAT LED TO THIS BOOK

Photographs courtesy of: William Strode Associates: p. 8; Photo Researchers: pp. 18 (Thomas R. Taylor), 20 (Hugh Spencer), 56 (Simon Fraser), 74 bottom (Hank Morgan); Panasonic: pp. 34, 51; AP/Wide World Photo: pp. 69, 129; Argonne National Laboratory: pp. 70, 99 (Nestor J. Zaluzec); CoolMax: p. 74 top; Battelle Memorial Institute: p. 91; University of Illinois: p. 96 (Matthew J. Zaluzec); NASA: p. 105; Oakridge National Laboratory: p. 124; Digital Instruments Inc.: p. 125.

Library of Congress Cataloging-in-Publication Data

Bortz, Fred.
Superstuff! : Materials that have changed our lives / Fred Bortz.
p. cm. — (A Venture book)
Summary: Explores the fields of materials science and materials engineering, including the development and technological properties of semi-conductors, superconductors, polymers, metals, alloys, ceramics, and glass.
ISBN 0-531-10887-2
1. Superconductors—Juvenile literature. [1. Superconductors.]
I. Title.
QC611.95.B67 1990
621.3—dc20
90-12565 CIP AC

Printed in the United States of America
6 5 4 3 2

CONTENTS

Prologue
"There's Lots of Superstuff Around Here!"

A magnetic resonance imaging (MRI) system is one of the most powerful new tools for physicians to diagnose the causes of illness. In an MRI test, the part of the patient's body to be examined is placed in the field of a very powerful magnet and exposed to pulses of radio waves. Hydrogen is weakly magnetic so the strong field of the magnet lines up the hydrogen atoms in the water of the patient's body. The radio waves knock the hydrogen atoms slightly out of alignment, and they begin to spin like tops, producing distinctive radio signals of their own.

Because the radio signals depend on what surrounds the hydrogen atoms that produce them, they can be analyzed by a computer to produce a picture of the inside of the body. MRI is similar to an X-ray CAT scan, in that it uses a computer to produce a picture of the inside of a person's body.

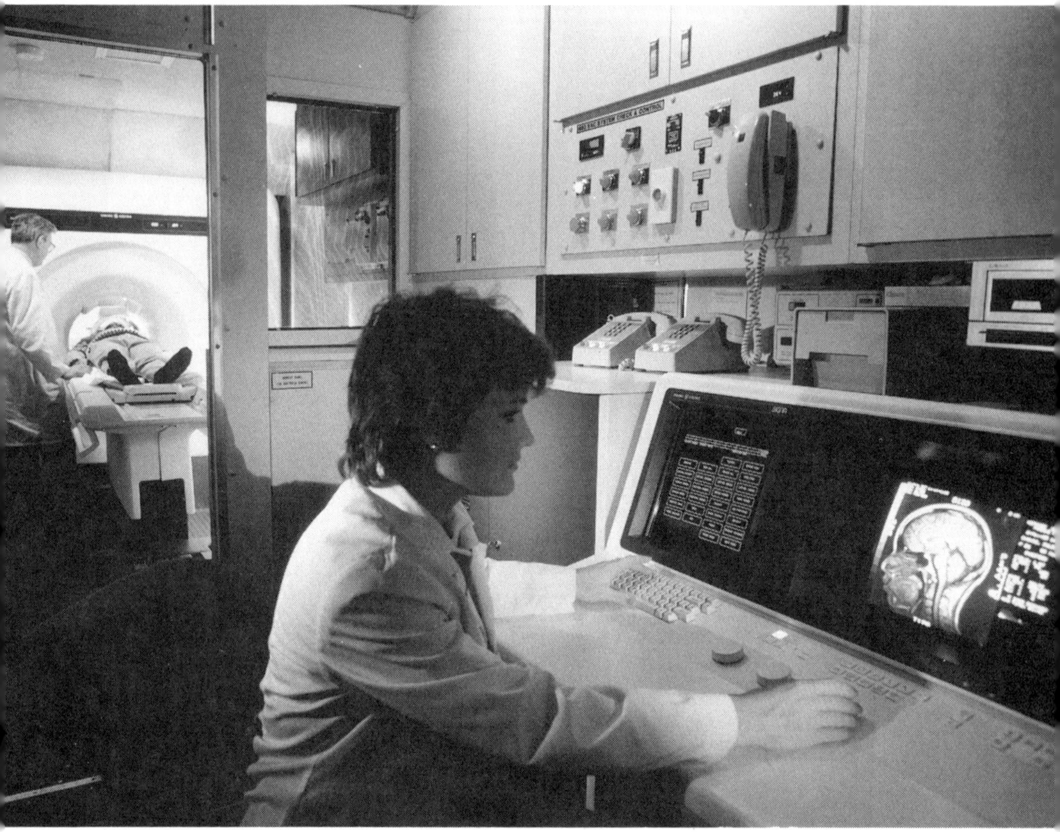

A magnetic resonance imaging (MRI) system used to diagnose causes of disease. MRI makes use of many advanced materials.

Many physicians prefer MRI to a CAT scan because X rays always damage a few body cells, and, as far as we know, MRI causes no harm at all. Furthermore, many physicians believe that MRI can make a much better image of brains and other internal organs than X rays can.

In the picture, the MRI magnet is the object shaped somewhat like a giant roll of toilet paper. The patient slides into the tube in the middle. This MRI magnet, like most, is an electromagnet made from superconducting wire; some MRI systems use permanent magnets made from newly discovered, powerful magnetic materials.

Superconductors have been in the news since 1987, when dramatic new discoveries opened up possibilities for their much wider use, but powerful superconducting magnets such as the ones used in MRI have been around for decades. Still, the materials in them are quite remarkable. And those aren't the only remarkable materials that make MRI possible. While the patient is in the magnet, the physician is in a room full of computers and video displays. The images on the displays are stored first on computer disk drives. Later, the images will be put on magnetic tape or an optical disk so they can be viewed again at any time. They can even be transmitted around the world by communication satellites and optical fiber communication systems. None of that could be happening without remarkable materials that did not exist until a few years ago.

So look around this world of superlatives that we live in, and think about what things are made of. Whether you speak of supercomputers, superadhesives, the Superconducting Super Collider, or the artificial grass in the Superdome, whether you are in a car, in school, in the hospital, or in your home, you can look around and say, "Wow, there's lots of superstuff around here!"

Chapter One
From Stone Tools to Superstuff

When people speak of "high technology," they usually refer to machines—spacecraft, computers, videocassette recorders, Walkman radio/tape players, car phones, medical imaging systems—all the wonderful inventions that have changed our lives for the better. Because of "high-tech" inventions, we have more knowledge, better health, more leisure activities, and the ability to communicate with more people than ever before.

This book looks at high technology in a very different, very basic way. It is not about how computers work, but what they are made of. It is not about the space program, but the heat-absorbing tiles that prevent the space shuttle from burning up on reentry. It's about the remarkable *materials* of today—semiconductors and superconductors, ceramics and composites, plastics and synthetic fabrics, special kinds of glass for fiberoptics, miracle alloys, superadhesives, and many other kinds of superstuff—and how they change our lives.

FROM THE STONE AGE TO THE BRONZE AGE

Now, you may think that the materials of today set us apart from people of earlier times, but that isn't really so. From before the dawn of recorded history, not only have people been making tools and machines—things we call "technology"—but they have also been discovering new materials. In fact, without the new materials, the new technologies of those times would never have come into being. So before we take a closer look at today's materials, let's go back to when technology started, more than two million years ago in a period we now call the *Stone Age.*

By today's standards, Stone Age technology is not very impressive. But two million years ago, humans discovered that hitting stones together could make them into sharp tools and weapons. That led to enormous changes. No longer did people simply try to survive in their environment; rather, they actively looked for ways to change the environment to their advantage.

Still, Stone Age people had to rely on the materials that nature gave them: skins, bones, teeth, and sinews from animals; fibers from plants; wood from trees; earth; and, of course, rocks and stone. That situation changed about ten thousand years ago when people discovered that they could make their own stonelike material by heating clay in a fire. Today, we would call that first human-made material a *ceramic.* Whatever those early people called it, they knew that they could form it into pots, bowls, jugs, and many other kinds of useful objects. This was the first new technology based on a material not found in nature, and it changed the way people lived.

About three thousand years after that, people discovered how to use *metals.* Metallurgy began with people's fascination with nuggets of copper and gold they found in the ground. Then came the discovery that metals did not chip like stone but rather could be pounded into useful shapes.

Other important discoveries followed. People learned how to melt metals and make objects of various shapes by pouring the molten metal into a mold. This process, called casting, could create metal tools and weapons in shapes that could not be made with stone. But since metal was rare and metal tools wore out more easily than those made of stone, people continued to use stone tools more than metal ones. That began to change with the discovery of two important metallurgical processes: *smelting* and *alloying.*

In smelting, a metal is produced by heating a substance called an *ore* in a furnace, sometimes with another substance. Ores are much more common in nature than pure metals, so the discovery of smelting greatly increased the amount of metal in the world. In alloying, two or more different metals are mixed together, usually in the molten state. The resulting material, called an *alloy,* is often more useful than the metals that went into it.

Besides increasing the amount of metal available for people's use, smelting also led to the discovery of new metals, such as tin. Tin, by itself, did not seem to be very useful, but because its ore was found with copper ore, people naturally tried mixing it with copper. The result was a very useful alloy called bronze. Bronze, discovered fifty-five hundred years ago, turned out to be the superstuff of the Stone Age. It was as strong as stone and, unlike stone, could be cast into many useful shapes. Because bronze tools had better shapes than stone tools and were just as strong and easier to make, stone tools were suddenly obsolete. With the creation of another material not found in nature, civilization changed dramatically. The Stone Age had ended and the *Bronze Age* had begun.

During the Bronze Age, people made many useful discoveries about metals. For example, they discovered that bronze became stronger when they beat it while it was hot. They also continued to discover new ores and new metals. Despite the discoveries, people found nothing to equal bronze for a long time.

THE IRON AGE

Finally, around 1900 B.C., another metallurgical breakthrough occurred: People discovered how to smelt iron from a very common substance. Iron was already known because it is the main component of some meteorites, but it was very rare.

After the discovery of how to smelt it, iron became very important because it became far more plentiful than bronze. People who knew how to smelt iron were able to make many more objects than before. Over the next seven centuries, as the new knowledge spread, civilization underwent a slow and gradual change from the Bronze Age to the *Iron Age.*

Iron became even more important following the accidental discovery that reheating it in a charcoal fire after smelting made it stronger. In fact, that reheating produced an iron-carbon alloy that we now know as steel. As centuries passed, steelmaking gradually improved; people learned how to make more steel and how to make it better. By the middle of the eighteenth century, they knew how to make large quantities of high-strength, high-quality steel. That knowledge, together with the invention of the steam engine, transformed the Iron Age into the *Industrial Revolution.*

So looking back on civilization from the Stone Age to the Bronze Age to the Iron Age to the Industrial Revolution of steam and steel, we see again and again the technological importance of the discovery of new materials. Today, we are in the midst of a period of changes more dramatic than ever before, a period driven by an explosion of new technology made possible by remarkable new materials.

Chapter Two
More Than Molecules

MATERIALS AND THEIR PROPERTIES

What makes a material hard or soft, magnetic or nonmagnetic, a good or poor conductor of heat or electricity? What permits it to pass or block light? to bounce or to shatter? When you first learn about matter, you are told that every substance is made of molecules and that each molecule is made up of atoms in particular combinations. The molecular makeup of a substance, you are told, determines its properties. Those statements are correct as far as they go, but they can be misleading. For example:

■ Diamond and graphite are both pure carbon. Yet diamond is clear and the hardest substance known, and graphite is so black and soft that it is mixed with clay to make pencil lead.

■ Bronze Age people discovered that beating a metal, especially when it is being heated in a fire, makes it tougher after it is cooled. Yet heating and beating doesn't change the molecules at all.

■ When sand is melted and cooled again, it becomes glass. Yet its molecules are still silicon dioxide, the same molecules that make up the mineral quartz.

You can begin to get an idea of how to explain those facts by thinking about water. As you probably know, a water molecule is made of three atoms, two hydrogen atoms and one oxygen atom combined together. If you cool water to a temperature of 0° Celsius (32° Fahrenheit), it will begin to freeze and eventually become ice. The temperature of the water and the ice are exactly the same and so are the molecules, yet the water is liquid and the ice is solid.

If you heat water to 100°C (212°F), it will begin to boil and eventually become steam. The temperature of the water and steam are exactly the same and so are the molecules, yet the water is liquid and the steam is gaseous.

Do those facts give you any ideas? Perhaps you know that it isn't the molecules or temperature that distinguishes ice from water at its freezing point; it's the way those molecules are arranged. In water, the molecules are fairly close together but are arranged at random. In ice, the molecules are about as far apart as before (actually just

Quartz crystals. Quartz is made up of silicon dioxide, as is sand and glass.

slightly farther apart on the average than in water), but they are arranged in a definite pattern. You can see evidence of this arrangement in snowflakes or frost *crystals* on cold window glass. Frost and snowflakes don't form as simply shapeless pieces of ice; rather, they develop intricate branches and patterns, with every branch at precisely the same angle. That angle is 60°, exactly one-sixth of a full circle, which is why snowflakes, when viewed up close, have beautiful and varied six-branched forms.

WHY MOLECULES FORM CRYSTALS

The shape of snowflakes is remarkable as well as beautiful. Imagine a tiny snowflake, about an eighth of an inch (one-third of a centimeter) across. Even though it seems small to you, it contains about a billion times as many water molecules as there are people on the Earth, and all those molecules are organized into a pattern. It is as if all the people on Earth and a billion other planets were organized into a giant marching band on a football field as big as our galaxy. The pattern formed by the billions of rows of billions of people on that field can be seen from a distant galaxy, even though the individual marchers would be far too small to notice.

If you think about how difficult it would be to organize that galactic marching band, you will soon ask how so many molecules can be organized into a crystal. After all, a molecule has no brain to follow instructions, and furthermore,

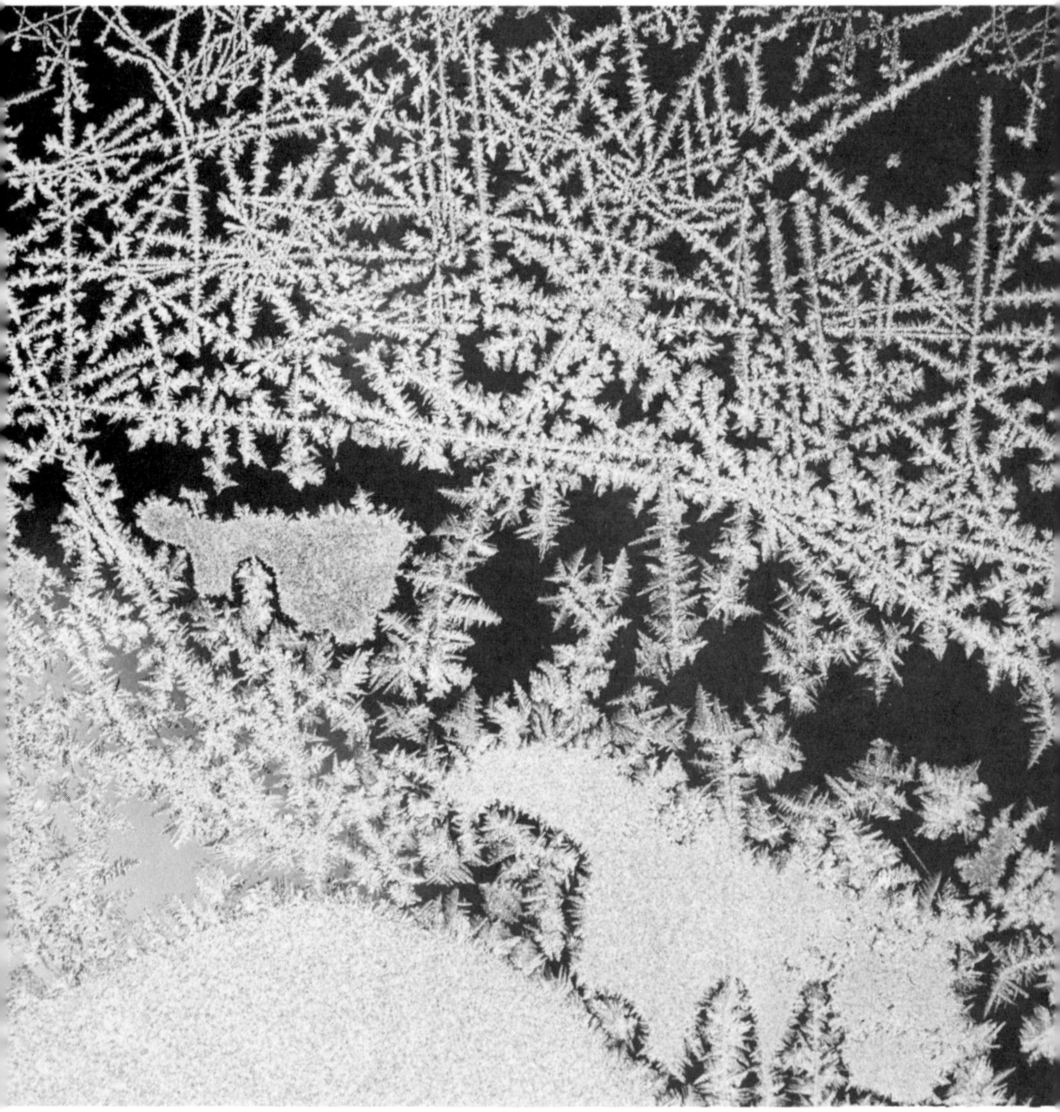

Frost on a windowpane displays the intricate branched patterns typical of ice crystals.

there is no molecule director with a baton to tell them what to do. Whatever happens to form a pattern must happen quite naturally. In fact, what happens is a lot like this:

Suppose you had a bag full of marbles, all the same size, and you began to put them in a box one at a time. The first marble would roll around and finally settle somewhere. The next marble would settle somewhere else. Little by little, the marbles would become more crowded, and soon they would form clusters. Soon the clusters would come together to form a layer of marbles. If you would jiggle the box a little, the marbles would shift around until there might be space for a few more marbles in the layer. Eventually, the layer would be full and new marbles would have to go into another layer on top of the first one. As you continued to add marbles, you would fill the box layer by layer.

When you do this with marbles, each layer has an interesting pattern. Each marble is surrounded by six others. If you were to draw lines from the center of one marble to the centers of the six that surround it, those lines would meet at 60° angles, just like the branches of a snowflake. If you drew all the center-to-center lines in a layer, you would create a pattern of triangles. In the middle of each triangle would be a low spot where a marble in the next layer would settle, so the pattern of one layer creates the pattern of the next, and so on. Thus quite naturally, just because the marbles are all the same size, they would form a regular arrangement.

Now if you did the same thing with little wooden cubes, you would get a different pattern. Pencils would give you another pattern, and bricks or department-store gift boxes would give yet other patterns. The pattern that you get would be different, depending on the object that you were using, but you could continue to add marbles or blocks or pencils or bricks or boxes by the billion until someone on a distant planet could see the pattern even though that person could not see the objects that make it up.

SAME MOLECULES, DIFFERENT CRYSTALS

So, crystals are quite natural, and most solid materials, like ice, are crystalline. But that doesn't explain why carbon crystals can be either diamond or graphite. To understand that, you have to think of molecules as something other than solid objects like marbles, bricks, or blocks.

Again, let's start with our familiar friend, water. A water molecule is made up of two atoms of hydrogen and one atom of oxygen. Like all molecules, it is held together by electrical forces between its atoms. Every atom is made up of smaller particles called protons, neutrons, and electrons. The protons and neutrons make up the core of the atom, called the nucleus. Each proton carries a positive electric charge, and the number of protons determines what kind of an atom it is. Hydrogen, the simplest atom, has one proton in its nucleus, and oxygen has eight. The nucleus of each atom is surrounded by a cloud of electrons, one for each proton. The electrons are negatively charged. Since opposite charges attract, the atom stays together.

The farther each electron gets from its nucleus, the weaker is the force holding it in place. So when two or more atoms come together, their outermost electrons may be nearly as close to the nucleus of another atom as they are to their own nucleus. As a result, the atoms can share or exchange electrons, and that binds them together into a molecule. In a water molecule, the atoms come together in a shape like a letter V with its arms spread apart (Figure 1). The oxygen atom is at the point of the V and the shared electrons are along the arms, spending more time close to the oxygen nucleus, with its eight protons, than to the hydrogen nucleus, with its single proton.

Thus, although the water molecule is electrically neutral, its point has a slight negative charge and the ends of arms of its V-shape have slight positive charges. When water molecules come together in an ice crystal, they don't just fit together like marbles; rather, they join so the positive end of one molecule meets the negative end of another. Just as electrical forces bind atoms together to form molecules, so do they hold molecules together in crystals. We can therefore think of a crystal as being made of building blocks that can be joined together in certain places but not in others.

Carbon atoms have four electrons that they can share when they join with other carbon

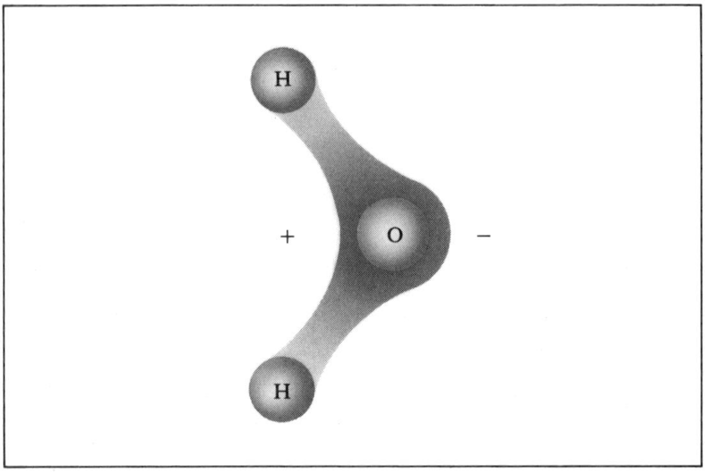

Figure 1. In a water molecule, two hydrogen atoms are joined to one oxygen in a V shape. The bond results from shared electrons. Because the oxygen nucleus has a much greater positive charge than the hydrogen nucleus, the shared electrons spend more time near the oxygen. The point of the V thus has a slight negative charge.

atoms. A common way of sharing electrons results in the graphite structure, as shown in Figure 2. As you can see, the carbon atoms form hexagons in layers, or "planes." The empty centers of the hexagons in one plane line up with the corners of the hexagons on the next. The carbon atoms form very strong bonds with three neighboring atoms in the hexagons. The fourth bond is shared between two other atoms, those directly above and below the carbon atom. This bond is very weak for two reasons: (1) the electron is shared among three atoms, and (2) the atoms sharing the electron are two layers apart.

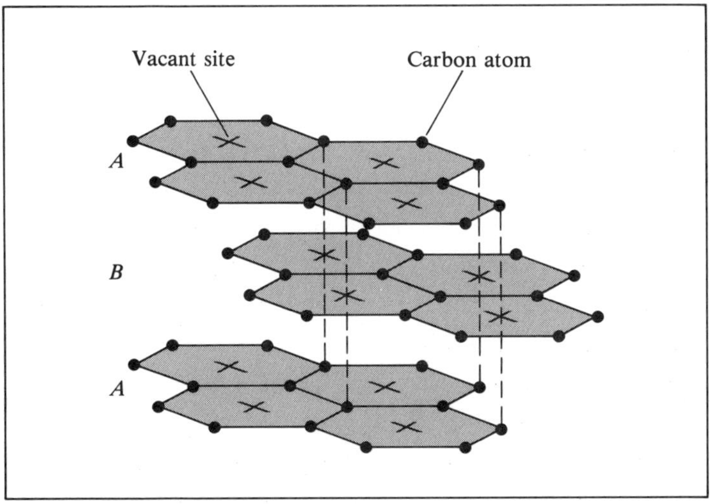

Figure 2. Graphite is a form of carbon in which the atoms form planes of hexagons with empty centers. It is easy for the planes to slip past one another. That makes graphite slippery and useful as a lubricant.

The weakness of one bond and the strength of the others give graphite unusual and useful properties. The crystal planes can slip past each other quite easily when the forces pulling on adjacent planes are in opposite directions, so graphite is slippery and can be used as a lubricant or in pencil "lead." However, thin graphite fibers can be very strong when the forces pulling on all planes are in the same direction, for those forces are opposed by the strong bonds within the hexagons. Later in this book, you will read about how that strength is used in *composite* materials.

In diamond, carbon atoms bond together not

in planes, but in a three-dimensional structure. As you can see in Figure 3, no matter which direction you push or pull on the diamond crystal, your force will be opposed by powerful bonds between carbon atoms. The strength of those bonds and the crystal structure make diamond the hardest material known.

THE IMPORTANCE OF IMPERFECTION

People usually notice that a material is crystalline because it has precise shapes, like the right-angle edges of rock salt, or it can be cut more easily along certain planes than others, like gemstones.

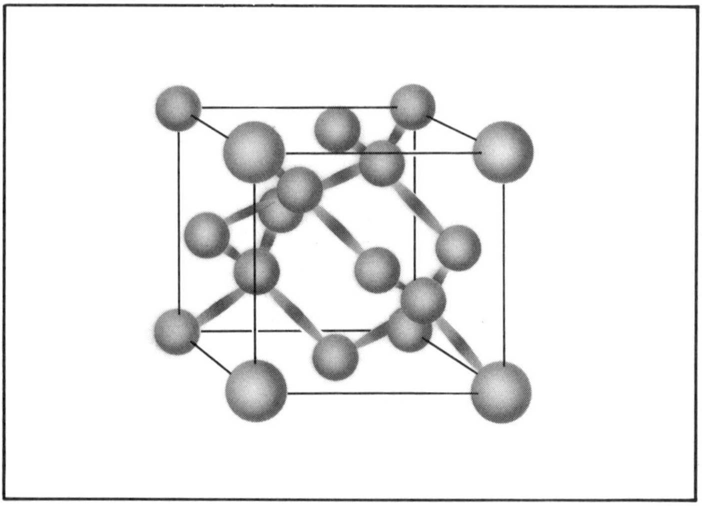

Figure 3. Diamond is a form of carbon in which each atom is bonded to four others in a three-dimensional structure. Unlike graphite, it is a very hard material because no matter which direction you push, your effort is opposed by a strong bond.

They sometimes think that glass is crystalline because it can be cut to look like a crystal (in fact, fine cut glassware is often called "crystal"), but glass, as you will discover later, is not a crystal at all. And although people rarely think of metals or ceramics as crystals, many important properties of those materials are determined by their crystalline nature—including their imperfections.

Nearly all crystals are imperfect. Any large crystal has its share of imperfections, or *defects.* With billions of billions of molecules in something as small as a snowflake, absolute perfection of any crystal is more than nature or the best of human technology can achieve. In fact, absolute perfection of a crystal is usually less desirable than controlled imperfection. Later on we'll discuss why that is so.

TYPES OF IMPERFECTIONS

Grains

Most solid substances are crystalline, but only a few of them ever form crystals large enough for people to notice. The rest are made up of many tiny crystal regions and are therefore called "polycrystalline." Each tiny crystal region is called a *grain.* Each grain has the same properties as any other, and if you were to examine the individual atoms in a grain, you would find them lined up in crystal planes with the same pattern and spacing as the atoms in any other grain. But if you were to compare two neighboring grains, you would find that their crystal planes did not line up with each other.

Point Defects

The pattern of atoms or molecules in a crystal is called a crystal *lattice,* and each position in that arrangement is called a lattice site. If you think of a crystal as a baseball stadium, the lattice is the evenly spaced arrangement of seats, and each seat is a lattice site. A perfect crystal is an arrangement of atoms on that lattice in a repeating pattern that is perfectly predictable. For example, a perfect sodium chloride crystal would be like a man in each even-numbered seat and a woman in each odd-numbered seat in one row and the opposite arrangement of the sexes in the next row.

Now if someone stays home, or if a lattice site is unoccupied, we have a *vacancy.* If someone sneaks into the stadium and squeezes in between two seats by standing on the armrest, or if an extra atom or molecule squeezes into the crystal where there is no lattice site for it, we have an *interstitial.* If a Martian steals someone's ticket and uses the seat, or if an atom of potassium is present where a sodium atom belongs, that is called an *impurity* (see Figure 4). Those are all different types of *point defects.*

A point defect affects the properties of the crystal not only at that lattice site, but also in the vicinity of that site. Atoms shift a bit to fill in some of the space left by a vacancy. An interstitial atom pushes nearby atoms a bit out of place, and those atoms push their neighbors, and so forth.

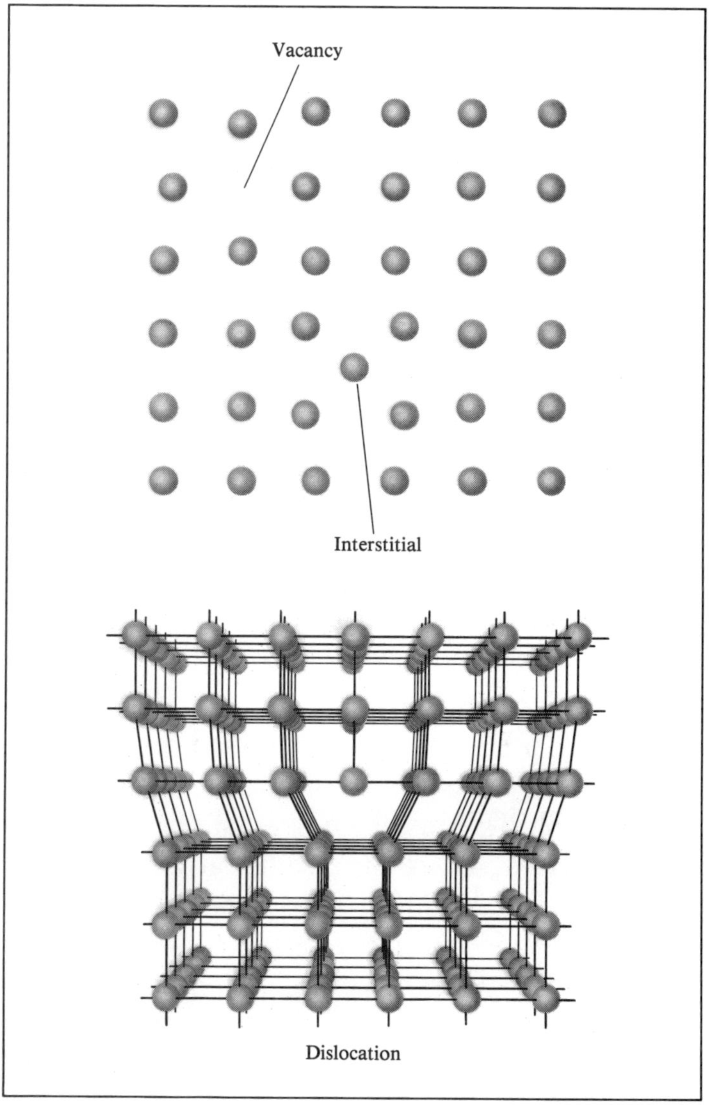

Figure 4. Many properties of materials are determined by imperfections in their crystal structure. Three common crystal defects are: (1) Vacancy. An atom is missing from its lattice site. (2) Interstitial. An atom is squeezed in between two lattice sites. (3) Dislocation. An entire plane of atoms is displaced.

An impurity can cause either of those effects, depending on whether it is larger or smaller than the atom it replaces. Or it can cause even more dramatic effects. Impurities in aluminum oxide give ruby and sapphire their distinctive colors, and carefully placed impurities turn silicon into transistors.

Line Defects

A *dislocation,* or line defect, occurs when the atomic planes of one section of a crystal lattice are slightly offset from another. The hardening of bronze when it is beaten is due largely to the movement of its dislocations until they reach a place in the crystal where their further movement is blocked.

Planar Defects

Two types of imperfections fall into the category of planar defects. One of these is a grain boundary, where two grains of a polycrystalline material come together. There the crystal lattice planes of the two grains are misaligned. The other type of planar defect occurs in crystals that have different atomic arrangements in neighboring planes. The planes normally follow one another in a repeating sequence, but a plane can occasionally be missing or out of order. When that happens, the crystal is said to have a stacking fault.

Volume Defects

A volume defect, usually called an *inclusion,* is a small region of the crystal that is entirely different from its surroundings. It may be an impurity, or it may be another crystalline form of the same substance. For example, diamonds often have inclusions of graphite, which greatly diminish their value as gems, but not their hardness. Such diamonds are called industrial diamonds because they are used to make industrial cutting and grinding tools. The word *bort* or *bortz* (no relation to the author) is used to describe low-quality industrial diamonds and diamond fragments in crushed or powdered form, which are used to make grinding or polishing tools.

NONCRYSTALLINE SOLIDS

Although all of the different types of superstuff described in this book are solids, not all of them are crystalline. An important class of materials is called *amorphous,* which means "without form." The arrangement of atoms and molecules in amorphous materials is irregular or random. There is no crystal lattice to dictate where an atom belongs. Instead of being like a stadium with a regular arrangement of people in their seats, an amorphous material is more like a beach where people spread their towels wherever they like and lie down. There may be clustering of people on a beach or of atoms in an amorphous material, but there is no fully predictable, regular arrangement. The molecules in an amorphous solid are arranged much like those in a liquid, but unlike a liquid, the material does not take the shape of its container.

The most important amorphous material is

also one of the most ancient materials made by humans. That material is *glass,* made from melted sand, and it's more important to technology today than it has ever been.

MICROSTRUCTURE AND SUPERSTUFF

Now you understand why the properties of a substance are determined by more than molecules. The arrangement of those molecules, the *microstructure* of a material, determines whether it is hard or soft, magnetic or nonmagnetic, a good or poor conductor of heat or electricity, transparent or opaque, bouncy or brittle. Microstructure is so important that its study has become the basis of a new science that developed out of the older sciences of metallurgy, chemistry, and physics during the second half of this century. That new field is known as *materials science.*

An important companion field to materials science is *materials engineering.* The materials engineer uses what people know about microstructure to design materials with properties that will be even more remarkable than the wonders of today.

Chapter Three

Electrical Conductivity and the Semiconductor Revolution

Did you ever hear someone say, "Wait, let me get my transistor," then go off to get a small portable radio? If you did, chances are that the person talking grew up in the 1950s, because that decade marked the beginning of a revolution in communications, computers, and consumer *electronics.* That revolution was made possible by a remarkable device, a bit smaller than a dime, called a *transistor.* Invented at Bell Laboratories by John Bardeen, Walter Brattain, and William Schockley, the transistor is based on the properties of a class of materials called *semiconductors.* Before transistors, electronic devices like radios, televisions, and computers used larger and much more power-hungry devices called vacuum tubes.

The average person's first contact with transistors was the hand-held transistor radio. These radios were a bit too thick and heavy to be put com-

The semiconductor revolution led to the development of the wealth of consumer electronics products on the market today.

fortably in a shirt pocket, but they were battery-powered and easily carried around. Suddenly radios appeared on the streets, on beaches, at ball parks. Teenagers hid their "transistors" under their pillows and listened to the radio when they were supposed to be going to sleep. Today, with radios as small and thin as credit cards and televisions that fit in your pocket, it may be hard to imagine what a breakthrough the invention of the transistor was. But the scientists of that time realized how revolutionary it was, and so did the people. In 1956, as most people were just learning about transistor radios, Bardeen, Brattain, and Schockley were awarded the Nobel Prize in physics for their invention of the transistor.

The transistor was only the beginning of the semiconductor revolution. The *integrated circuit (IC),* with many transistors on a tiny chip of silicon, followed about ten years later, opening the way to more-advanced computers. The first people to benefit from those advances were business people, scientists, communications companies, and manufacturers. But just as the transistor quickly found its way into the "consumer market"—the market for items sold in large numbers to ordinary people—the IC soon appeared in such everyday items as home appliances. More than that, calculators and computers became consumer items as well, and the initials PC, for personal computer, became part of the world's vocabulary.

Other products, such as VCRs (videocassette

recorders) and CD (compact disk) players, are among the most recent breakthroughs in consumer electronics. The CD player is a particularly significant development of the semiconductor revolution, for not only does it use an IC-based computer and other semiconductor devices to create the most lifelike sounds ever reproduced electronically, but it also uses a semiconductor laser about the size of the first commercial transistors and a photodetector, a light-sensitive semiconductor, to produce the electric signal that the computer needs to produce that sound.

How do those semiconductor devices work? Before we can answer that, we have to take a journey into the microstructure of solids to understand what makes them electrical insulators, conductors, or semiconductors. It's a fairly long journey, but there is a lot of interesting learning along the way. So be patient; we'll get back to semiconductors a few pages from now.

BONDS, BANDS, AND ENERGY LEVELS

The electrical properties of a substance are determined in large part by the bonds between the atoms or molecules that the substance is made of. As you read in the last chapter, these bonds are created by the sharing of some of the atoms' electrons, the negatively charged particles that surround the positively charged nuclei (plural of "nucleus"). When the atoms come close enough together to bond, their outermost electrons are shared, forming bonds which can be of three major types, called *ionic, covalent,* and *metallic bonds.* (In certain materials, the bonds have characteristics that are a mixture of two of these types.) To understand the different types of bonds, it is important to understand the concept of *energy levels.*

Energy Levels

Imagine that you are building an atom, starting with the nucleus and adding electrons one at a time. Because of certain laws of nature that are described in a branch of physics called *quantum mechanics* or *quantum theory,* this process can be thought of as if the electrons are being dropped onto an uneven flight of steps, with each step having room for only a certain number of electrons. These steps are called energy levels because it would take a certain amount of energy to remove the electron from the nucleus and take it to the top of the stairs. Each energy level is said to contain one or more "shells" of electrons. The energy needed to remove an electron from the lowest step is 1 unit. The energy needed to remove an electron from the higher steps decreases to ½ unit on the first step above the bottom to ⅓ to ¼, and so on. In terms of decimal numbers, the energy levels are (to the nearest thousandth of a unit) 1.000, 0.500, 0.333, 0.250, 0.200, 0.167, 0.143, and so on.

Ionic bonds

Now let's return to the idea of building up an atom electron by electron. As you add each electron, it goes into the lowest energy level available.

The shells and levels gradually fill until you have no more electrons to add.

This picture is fine for one atom, but when two atoms come close together, things change a bit because nature tends to favor filled shells. That means that an electron in a nearly empty shell is loosely bound to its atom. Such an electron may be easily captured by a nearly full shell of a neighboring atom of a different type. So instead of being electrically neutral, the atoms each have a charge. When that happens, they are called ions. The atom that captures the electron becomes a negative ion and the one that donates the electron becomes a positive ion. The oppositely charged ions are bound together by their electric attraction, and the bond between them is called ionic.

Ordinary salt, sodium chloride, is the classic example of ionic bonding. The electronic structure of a free sodium atom is filled shells plus one electron. A free chlorine atom has seven electrons in its outermost shell, one short of the number needed to fill it. When those atoms come together, the sodium donates its loosely bound electron to fill the chlorine's outer shell. Both atoms become ions and combine through a powerful ionic bond.

Covalent Bonds

Covalent bonding occurs when two atoms share electrons in their outer shells so that all shells are full. A water molecule is held together by two covalent bonds. Its oxygen atom has six electrons in its outer shell, which needs eight to be full. The shell holding the single electron in a hydrogen atom would be full if it had two electrons. So when two hydrogen atoms meet one oxygen, the oxygen forms one covalent bond with each hydrogen. In each bond, the oxygen atom shares one of its outer electrons with the hydrogen, while the hydrogen shares its electron with the oxygen. As a result of this sharing, the oxygen's outer shell has the eight electrons it needs to be full (its own six plus one from each of the two hydrogens), while each hydrogen has the two electrons (its own plus one shared by the oxygen) needed to fill its shell.

The outer electrons are called valence electrons, and the bond formed by sharing valence electrons is called covalent. Covalent bonds can be single, where each atom shares one electron with another, or multiple, where atoms share more than one electron to form the bond. In water, there are two single covalent bonds. In oxygen gas, the oxygen atoms pair off to form oxygen molecules consisting of two atoms of oxygen bound together by a double covalent bond. Nitrogen atoms, with five of the eight electrons needed to fill their outer shells, also combine as pairs. A triple covalent bond joins the two atoms of a nitrogen molecule.

The most interesting covalent bonding takes place in substances that have atoms with half of the eight electrons needed to fill their outer shells, such as carbon and silicon. This enables them to combine in many different ways with many different atoms, with single, double, and triple bonds. In diamond, for example, each carbon atom forms a single covalent bond with four others in the pattern you saw in the last chapter.

Metallic Bonds

When we imagined building up an atom an electron at a time, our picture made no allowance for other atoms. But when we discussed ionic and covalent bonds, we realized that other atoms make a difference. To understand ionic and covalent bonding, we could have imagined building up a solid by starting with a huge number of nuclei on a crystal lattice and adding an even bigger number of electrons one at a time, starting at the lowest energy levels and building up. Had we done that, we would have discovered that in addition to the energy levels for electrons associated with individual atoms, there are also energy levels for electrons shared between two atoms, for electrons shared among three atoms, and, in fact, for electrons shared among any number of atoms.

There would be so many energy levels of those new types that we could often not distinguish among their energies. Scientists have solved that problem using quantum theory and have found that the energy levels still form an uneven staircase, but that some of the steps are not single, sharp energy levels. Those steps, which they call *energy bands,* are more like bins that can hold enormous numbers of electrons. The electrons in the bands can have any energy in that band, just as a ball in a filled bin can be found anywhere between the top and bottom of the bin. But between the top of one band and the bottom of the next, there are ranges of energy in which there are no allowed electron energy levels, just as there would be no place to capture a ball between two bins.

Two particularly important bands of energy are called the valence band and the conduction band. The valence band contains energy levels for electrons shared between two atoms, such as those that make up a covalent bond. The conduction band contains energy levels for electrons that are shared among all the atoms. The valence band is always at a lower energy than the conduction band. Many properties of materials depend on how many electrons are in each band, and that, in turn, depends on how many single-atom energy levels are below the valence band. For example, in substances that form ionic bonds, there is enough room in the single-atom energy levels below the bands for all the electrons. Thus, both the valence and conduction bands are usually empty in ionic materials.

In substances with covalent bonding, the single-atom energy levels below the valence band are all full and the bonding electrons occupy the energy levels at the bottom of the valence band. The rest of that band is empty, as is the conduction band.

In a metal, however, the single-atom levels are all full, and so is the valence band. Many electrons occupy energy levels in the conduction band. Those are the energy levels that quantum theory says are for electrons shared by all of the atoms in the crystal lattice. They are not bound to any atom in particular, so they can move easily from one end of the metal to the other. Metals can thus be thought of as crystals of positive ions held together by being immersed in a negatively charged fluid of electrons.

ELECTRICAL CONDUCTIVITY

Now we are ready to discuss a very important property of matter, electrical *conductivity.* An electric current is the flow of electric charge from one point to another. The flowing charge can be ions or electrons, and it is driven by an electrical pressure, or electromotive force, that can cause positive ions to move in one direction and negative ions or electrons in the other. Batteries or generators can produce an electromotive force, which is commonly called voltage since its strength is measured in volts. The electrical conductivity of a material is a measure of how easily an electromotive force produces a current through it.

Good conductors have high conductivity; good *insulators* have very low conductivity. Using what you know about the three types of bonds between atoms, can you predict what kinds of materials would be good electrical conductors and what would be good insulators?

Let's start with ionically bonded materials like salt. Even though every lattice site in a crystal of salt has an electrically charged ion, solid salt is an excellent insulator because none of the ions is free to move. In addition, salt has no electrons in its conduction band; no electrons flow through it because every electron is tightly bound to an ion as part of a complete shell. What would happen if you melted the salt or dissolved it in water? Then, although the electrons would still be tightly bound to ions, the ions themselves would be free to move, and thus a current would flow. The positive sodium ions would move to the negative end of the battery or generator, while the negative chloride ions would move to the positive end. So although solid salt is an excellent insulator, molten salt and salt solutions are good conductors. The same is true of any ionic substance.

What about a covalently bonded substance? Some are semiconductors, and you will read more about them soon, but most are excellent insulators. Sand, for example, is a compound of one silicon atom and two oxygen atoms, with double covalent bonds joining each oxygen atom to the silicon. It is an excellent insulator because it has no ions and no electrons in its conduction band. Even its most weakly bound electrons are part of covalent bonds and are thus held in completed shells. Even when melted, sand remains an insulator because its molecules do not break apart into ions. Like most covalent substances, sand does not dissolve easily; but even if you could dissolve it, it would have no way to conduct electricity.

Why then, since water is covalent and therefore a good insulator, are you warned to avoid standing in it while touching something electric? The reason is that water is a very good solvent of ionic substances like salt. So although absolutely pure water hardly conducts electricity at all, most household and natural water has a reasonable amount of ionic material dissolved in it, and, as you just read, ionic solutions conduct electricity very well.

The electrical conductivity of the last class of bonds, metallic bonds, is probably the simplest one for you to predict. Before you opened this book, you probably knew that metals were good conductors of electricity, but now you know why. A metal has lots of electrons that are moving freely through it all the time. When a metal wire is connected to a battery or generator, those electrons quickly move toward the positive end, while new electrons replace them by entering the metal from the negative end. Thus it is no wonder that the energy band that contains those mobile electrons is called the conduction band!

But things are not as simple as they seem. Consider this question: why are some metals better conductors than others? As you might expect, some metals have more conduction electrons than others just because their atoms are different. But that is far from the complete answer. The rate at which current flows through a wire depends not only on how many electrons are moving, but also on how long it takes an electron to travel from one end of the wire to the other. As the electron travels through the wire, it continually bumps into things, and that usually slows it down. The voltage causes it to pick up speed again, so if it doesn't have too many collisions, it can get through quite rapidly. Among the things that it collides with are other electrons, atoms, and lattice defects. That means that conductivity in metals can be as varied as their microstructure, which makes it an interesting and often complicated subject.

For example, small amounts of impurities can make a big difference in a metal's conductivity, especially if the impurities have nearly complete outer shells to capture electrons that would otherwise be free. But if you think impurities are important in conductors, you haven't seen anything until you see what they can do for semiconductors!

SEMICONDUCTORS

Semiconductors are important not because of how well or poorly they conduct electricity, but rather because of the ways in which their conductivity can be modified. By making careful changes in their microstructure, people can create materials with very interesting and useful electrical properties. The most commonly used semiconductor is silicon.

A crystal of pure silicon is an excellent insulator. Each silicon atom has four valence electrons and is bonded to four other silicon atoms by single covalent bonds, creating a filled shell with eight electrons. It would not conduct electricity at all if all of those valence electrons stayed in the valence band all the time. However, valence electrons occasionally become conduction electrons. To understand why, we have to change our mental picture of silicon a little bit by considering the effect of temperature.

The temperature of an object is a measure of how much energy its atoms and molecules have. The hotter it is, the more those atoms and molecules jiggle around. At any instant, some are jiggling a lot and some are jiggling a little. They can collide and exchange energy with one another. That bumps some electrons in the valence band into the conduction band. This creates two kinds of charged particles in the crystal, negatively charged conduction electrons and positively charged silicon ions at lattice sites.

The ions have "holes" in their outer shells of electrons. A hole can easily capture an electron from a neighboring atom, completing the outer shell and creating a hole in the next lattice site. The hole moves from site to site in this way (Figure 5). For every electron in the conduction band, there is a positively charged hole in the lattice, and both move freely throughout the crystal. Eventually, a conduction electron encounters a hole and is captured by the electron-hungry ion. The conduction electron and the hole "recombine" and disappear.

The number of "charge carriers"—conduction electrons and holes—in the lattice depends on the temperature of the semiconductor. The

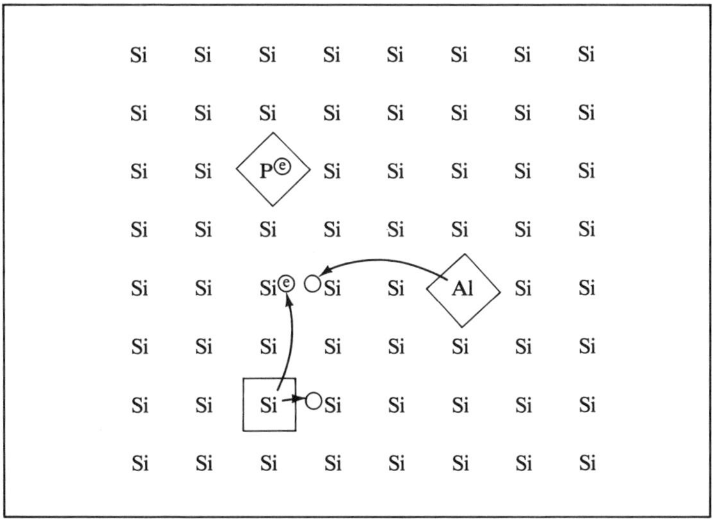

Figure 5. Electrical conductivity in a silicon semiconductor. Atom in square box jiggles, which breaks loose one of its electrons (e). Electron has moved through lattice. Hole left behind (open circle) captures an electron from the atom to its right, so hole moves to atom whose electron was captured.

In diamond-shaped boxes are two impurity atoms. Phosphorus (P) has an extra electron to add to the lattice. Aluminum atom (Al) introduces a hole that has moved two silicon atoms to the left, where it will capture the electron that broke loose from silicon atom in the square box.

higher the temperature, the more frequently electron-hole pairs are formed, and thus the greater is the conductivity of that semiconductor.

Now imagine that the crystal of silicon has some impurity atoms at some of its lattice sites. If the impurity is phosphorous, which has five electrons in its outer shell, covalent bonding between it and the silicon atoms on its neighboring lattice sites fills all the electron shells and leaves one electron left over. Since the valence band is full, that extra electron goes into the conduction band. Although there is a positive ion left behind, this time there is no hole. The more phosphorous, the more electrons are in the conduction band.

If the impurity is aluminum, which has only three electrons in its outer shell, covalent bonding between it and its neighboring silicon atoms leaves it and one of the neighbors with only seven of the eight electrons needed to complete the shell. The incomplete silicon atom can complete the shell by capturing an electron from another neighbor, which then captures an electron from another neighbor, and so on. Thus the addition of an aluminum atom and the capture of a neighboring electron create a negative ion that stays in its place on the lattice and a hole that moves around. This time there is a hole produced without a conduction electron. The more aluminum atoms, the more holes in the lattice.

In both cases, adding an impurity increases the number of charge carriers in the semiconductors, and thus its conductivity. These deliberately impure semiconductors are of two types, called "n-type" and "p-type," according to whether the impurity adds negative (n) or positive (p) carriers. When an impurity is added in a controlled way, the process is called *doping* and the impurity is called the "dopant." Doping silicon with phosphorous or arsenic adds negative conduction electrons, so it creates an n-type semiconductor. Doping silicon with aluminum or boron creates a p-type semiconductor. Other dopant atoms may be used, but these are the most common.

BASIC SEMICONDUCTING DEVICES

When you put n-type and p-type semiconductors together, you can make some interesting devices. The simplest of these is a diode, which is made by doping a semiconductor so that it is half p-type and half n-type. The region where they come together is called a p-n junction.

Diodes

A diode permits current to flow in one direction but blocks it in the other. Let's look at its charge carriers to see how it does its job (Figure 6). When the diode is not connected to a battery or generator, holes wander around in the p-type region, and conduction electrons wander around in the n-type region. Near the junction, electrons and holes recombine, so there are not very many carriers there.

Now suppose you attach a battery to the diode so that its positive terminal is connected to the p-type region and its negative terminal to the n-type region. In the n-type region, electrons leave the negative end of the battery and are driven to the junction by the battery's electromotive force. Meanwhile, in the p-type region, holes move away from the positive terminal of the battery toward the junction where they meet electrons. The holes

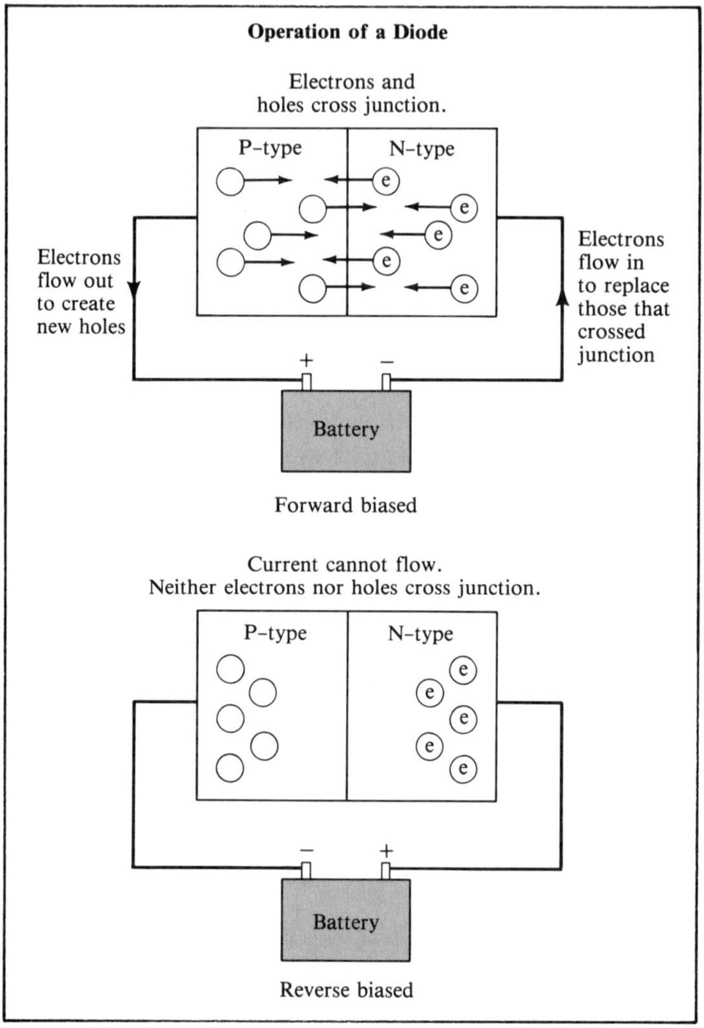

Figure 6. A semiconductor diode is formed at the junction of an n-type region and a p-type region of a semiconductor. It permits electrical current to flow through it in only one direction. When it is "forward biased" (top), the negative terminal of a source of direct current such as a battery is connected to the n-type region and the positive terminal to the p-type region. In "reverse-biased" diode (bottom), terminals are connected in the opposite fashion.

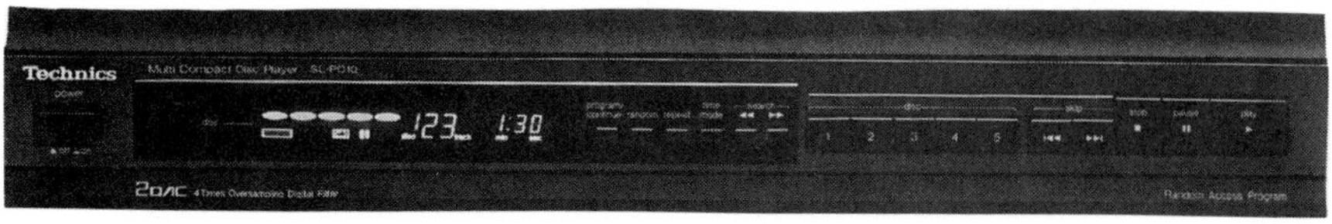

A numeric display created by light-emitting diodes (LEDs)

and electrons recombine, but new carriers replace them. In the n-type region, new electrons flow in from the battery's negative terminal, replacing the ones lost at the junction. In the p-type region, when electron-hole pairs are created, the electrons tend to flow out of the semiconductor and into the battery's positive terminal, leaving behind holes to replace the ones that recombined at the junction. Thus, current continues to flow. For that reason, when the diode is connected in this way—n-type region to negative, p-type region to positive—it is said to be "forward biased."

If you reverse the battery terminals, no current will flow because the electrons and holes, instead of being pushed toward the junction, are drawn away from it. In this "reverse biased" state, without charge carriers at the junction, no current can flow through the diode. Therefore a diode is like a one-way valve for electric current. That turns out to be very important because many electronic devices work only with one-way current, also known as direct current, or DC. Since household electricity is two-way alternating current, or AC, diodes are used to let AC pass when it is flowing in the desired direction and to block it when it would normally flow in the opposite direction. A device used in this way—to transform AC to DC—is called a rectifier.

Transistors

A transistor is two diodes back-to-back or three semiconductor regions that form an n-p-n or a p-n-p sandwich. To understand what it can do, let's look at the n-p-n type. (If you would like, you can see how well you understand this by trying to rewrite it for a p-n-p transistor.)

Figure 7 shows the three regions of the transistor labeled not only by their type (n-p-n) but also by their function (emitter-base-collector). There are two DC voltage sources, one large and constant, the other small and variable. The negative end of the larger voltage is connected to the emitter, and its positive end is connected to the collector. This causes the emitter-base junction to be forward-biased and the base-collector junction to be reverse-biased. If that were the only voltage source, no current would flow since the base-collector junction would stop it. (If you reversed the battery, you would reverse the biases of both junctions, and again no current would flow because the base-emitter junction would block it.)

The second voltage source is connected across the base-collector junction with its negative end connected to the collector and its positive end to the base. If that were the only voltage source, the base-collector junction would be forward-biased,

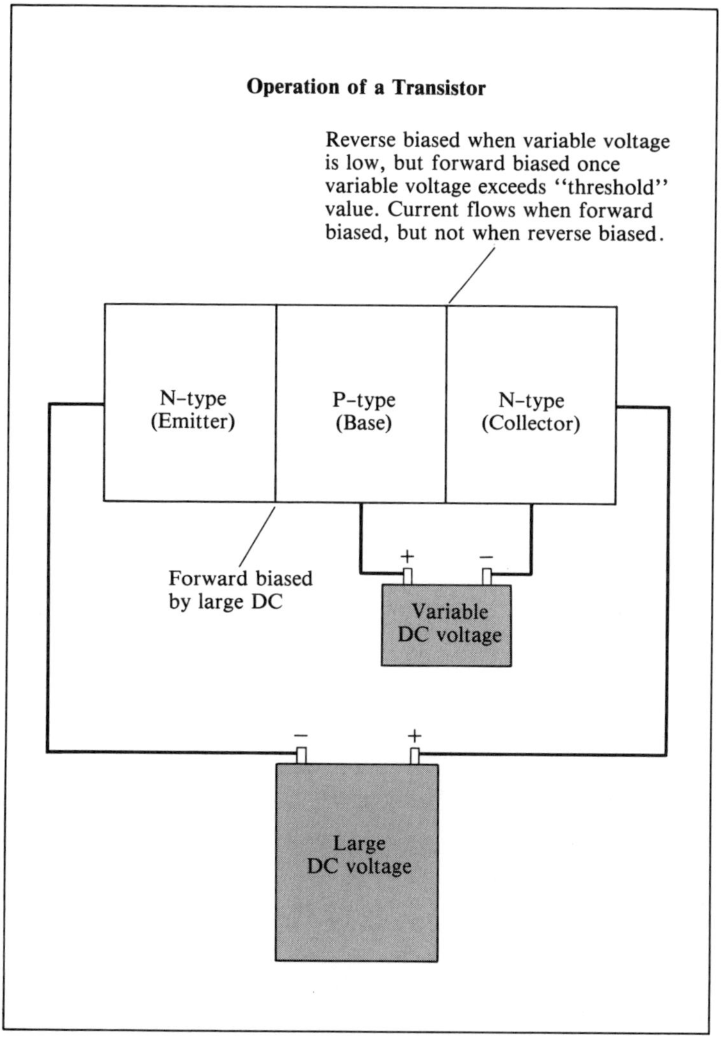

Figure 7. A transistor is formed from two diodes back-to-back in an n-p-n or p-n-p sandwich. By operating near the threshold level, a small change in the voltage of the variable DC source can cause a large change in the current through the transistor. Thus a transistor can be an amplifier or a controllable electrical switch.

but because of the first voltage source, it may not be. To understand what can happen, let's assume that we have a dial we can turn that gradually increases the voltage of this second source. If we started out with the small voltage at zero, the base-collector junction would be reverse-biased by the large voltage source, so there would be no current. Gradually, as we turned the dial, we would reduce the amount of reverse bias until we finally overcame it. At that point, which we can call the threshold voltage, current would begin to flow through the transistor. A very small additional change in the dial would lead to a very large change in current.

Why is a transistor's behavior valuable? In fact, there are two different things that it makes possible. In the simplest sense, a transistor can be used as a controllable on-off switch. Keep its base-collector voltage low and it stops the current; raise its voltage above the threshold and the current flows. Used in this way, the transistor can be part of an automatic control system. As a simple example, if the base-collector voltage source is a thermometer that produces a greater voltage when the temperature rises than at a certain threshold temperature, it can turn on an air conditioning system, a sprinkler, or a warning signal. This on-off switch behavior makes transistors useful in digital computers, where all of the information is stored as a pattern of on-off signals that represent a code of ones and zeros.

Another way the transistor can be used is as an amplifier. That is one of its main functions in a radio. Because the flow of current is so sensitive to the base-collector voltage, a small variation in that voltage can lead to a large change in current. A radio works by picking up an electromagnetic wave and turning it into a very small voltage. The amplifier makes it possible for that small voltage to be turned into a very big sound.

Integrated Circuits (ICs)

The first transistors were about the size of a dime and were connected by ordinary wires, but it was soon discovered that a transistor could be made much smaller. As long as the transistor did not have to carry much current, it could be tiny indeed. Soon people began to make several transistors and their connections on a single piece of silicon and called the device an *integrated circuit (IC)*. From the first ICs containing a dozen or so parts, people progressed steadily and rapidly to tinier transistors and more complicated integrated circuits. Today, the main working parts of powerful computers are fingernail-size chips of silicon containing millions of interconnected transistors.

SEMICONDUCTORS AND LIGHT

Light Producers

Not all important semiconductor devices are made of silicon because there are some materials that work better or have properties that silicon does not. An increasingly important semiconducting material is gallium arsenide, which is used, among other things, to make light-emitting

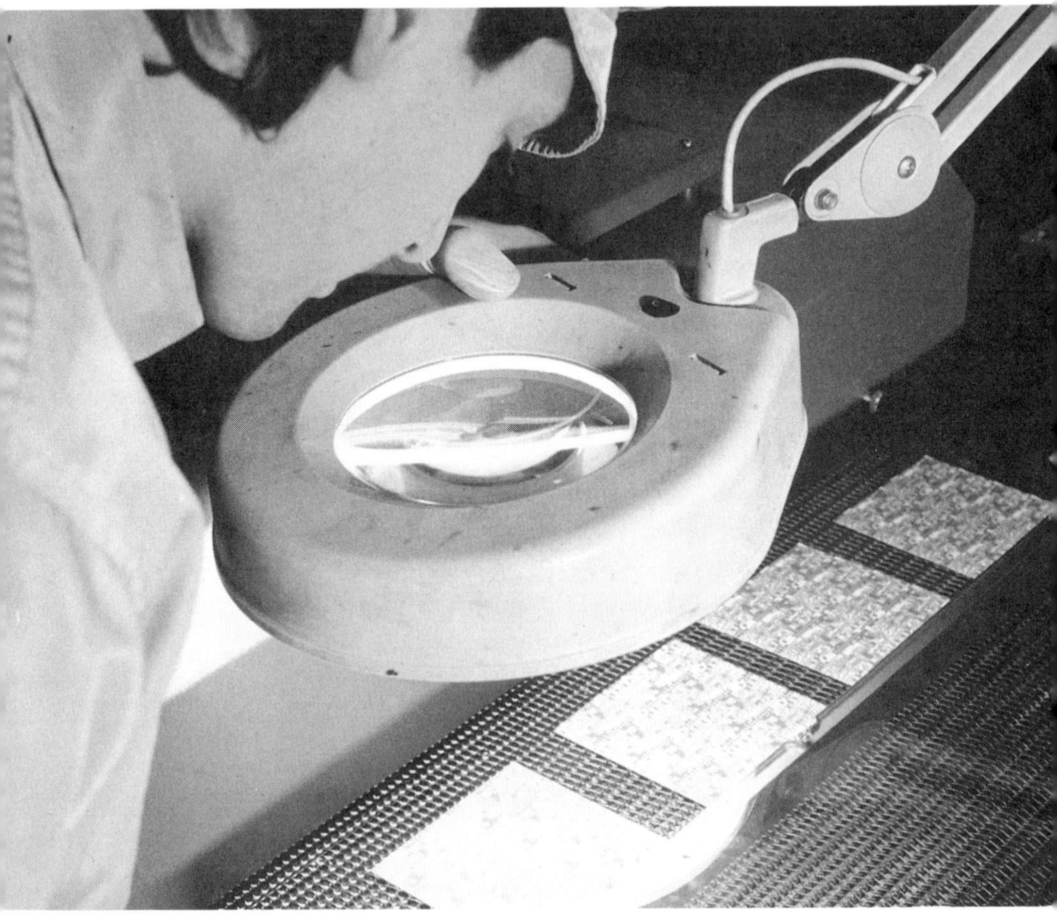

Manufacture of electronic hybrid circuits. Hybrid circuits consist of interconnected transistors, resistors, capacitors, and other electronic components. Here a worker checks the quality of the metallic conducting pathways.

diodes (LEDs). You have no doubt seen a group of LEDs, laid out in a rectangular pattern, that light up to form letters, numbers, or some electronically produced message, such as the time of day.

LEDs produce their light during the recombination of electrons and holes at their p-n junction. When an electron makes a transition from one energy level to a lower one, in this case from the conduction band to the valence band, the lost energy shows up as a particle called a photon. Light and other forms of electromagnetic energy are made of photons. The color of the light or the wavelength of the electromagnetic wave produced depends on the difference between the energy levels. Related to LEDs are laser diodes, tiny semiconductor devices that produce light with its waves all in step, making it more powerful. The laser in a CD player is made from a small piece of semiconducting material.

Light Detectors

Just as electrons produce photons when they lose energy, so they can absorb photons and jump from one energy level to a higher one. Thus semiconductors can be used as "electric eyes" to detect light. There's a semiconductor device like that to pick up the reflected laser beam in a CD player. But semiconductors can do more than that; if their energy bands are properly spaced, they can absorb photons of other electromagnetic waves. Semiconductors like mercury telluride or cadmium telluride respond to photons of infrared light—so-called heat waves. They are used to make devices that enable people to see warm objects in the dark or hidden by foliage. They can be lifesavers for soldiers or police.

Photonics

We've come a long way from the transistor. Today's young people will never be caught saying anything that shows their age like that "Wait, let me get my transistor" person said. After all, they grew up with CDs, PCs, VCRs, and all those wonderful electronic things. They'll never show their age when it's time to go to the electronics store.

If that's what you think, then you're wrong. Electronic devices may be on their way out, and photonic devices may be on their way in. Where electronic devices need semiconductors to carry their signals, photonic devices, based on light, will use transparent materials like glass, plastics, or nothing at all. While *electronics* is certainly "faster than a speeding bullet," *photonics* is as fast as light itself.

Just as no one could have predicted today's electronic wonders when the first transistor radios appeared, no one can predict what wonders will await us in the photonics store in 2025. Whatever they are, one thing is certain: photonic devices will require more advances in materials like semiconductors, glass, and plastic and inventive people who know how to use them.

Chapter Four
Superconductors: Hot Stuff!

You may remember that one of the hottest news stories of 1987 was the development of a new class of materials called "high-temperature superconductors." Superconductivity, a property of some materials at very low temperatures, had been known since 1911. Professor Heike Kamerlingh Onnes of the University of Leyden in the Netherlands discovered it while studying how the resistance of the metal mercury varied as it was cooled closer and closer to *absolute zero.*

(Absolute zero is approximately –273° Celsius, or –459° Fahrenheit. However, scientists usually measure low temperatures in what is called the Kelvin scale, whose unit of measurement is equal to the Celsius degree. The important difference between the Kelvin and Celsius scales is that the Kelvin zero point—0 Kelvin, or 0K—is absolute zero rather than the freezing point of water.)

Kamerlingh Onnes had expected the resistance of mercury to decrease as he cooled it. Study of the resistance of metals at normal temperatures showed that lowering the temperature decreased resistivity. Kamerlingh Onnes and other scientists reasoned that the decrease in resistivity was due to the fact that atoms vibrate less vigorously in a metal at a low temperature than at a high one, and are therefore less likely to collide with the flowing electrons. But no one had studied temperatures as extreme as those that Kamerlingh Onnes could create in his laboratory—temperatures low enough to turn all the gases in the atmosphere to liquids. So scientists looked to that laboratory to tell them whether the steady decrease in resistance with cooling would continue all the way down to the lowest temperatures possible.

As the temperature decreased below the temperature of liquid nitrogen (–196°C, –321°F, or 77K), the resistivity of mercury continued its steady, gradual drop, as did the resistivity of other metals and alloys that Kamerlingh Onnes studied. But at 4.2K, just barely below the boiling temperature of liquid helium, the resistance of his thin mercury thread dropped abruptly to zero. It was an astonishing discovery. The atoms of the metal still vibrated, but the electrons flowed past them and between them without losing any energy.

There were two possible explanations for what had happened. Either the electrons had stopped colliding with the atoms or their collisions had become perfectly elastic; that is, the electrons left the collisions with exactly as much energy as they had when they hit the atoms in the first place. Neither idea made sense, but the experimental facts said one of them must be true.

Kamerlingh Onnes tested his discovery in many ways. He tried the same wire again and again. He tried different shapes of wires. No matter what, superconductivity always began when he cooled the mercury to or below 4.2K. He tried other metals, and found that some of them, too, became superconductors—tin at 3.8K and lead at about 6K, for example. He made loops of superconducting wire and started current flowing in them by passing the pole of a magnet through the center of the loop. Kamerlingh Onnes of course used no continuous source of voltage, such as a battery or a generator; therefore, under normal conditions—that is, at any point above the critical temperature—the currents would have stopped flowing immediately. However, he came back hours later to find the currents not only still flowing but flowing as strongly as before. Then he raised the temperature above the critical temperature and the currents promptly died away.

Kamerlingh Onnes's careful work demonstrated that superconductivity was real, although it would be nearly a half century before anyone could explain it. In 1913, scientists honored Kamerlingh Onnes with the Nobel Prize for physics, citing not only his discovery of superconductivity but also his pioneering work that led "amongst other things, to the preparation of liquid helium." None of those present at the awards ceremony knew that superconductivity would be a hot topic seventy-five years after its discovery. Nor did they know that it held enough fascination and promise to produce two more Nobel Prizes.

In the years following Kamerlingh Onnes's discovery, many scientists searched for materials that became superconducting at higher temperatures, while others searched for a theory that explained the phenomenon. Still other scientists and engineers imagined what they might build with such materials, like superconducting wire coiled into powerful electromagnets that never got hot.

Progress in the theory was slow; it required first the development of quantum mechanics, then the understanding of energy bands for electrons in solids. Finally, in 1957 John Bardeen (who one year earlier, with his colleagues Walter Brattain and William Schockley, had received the Nobel Prize in physics for their invention of the transistor), Leon N. Cooper, and John Robert Schrieffer wrote a famous article that explained what happened to electrons in superconductors. Their theory, known as the *BCS theory,* after the initials of their last names, explained that each electron in the conduction band has a partner, with which it forms a "Cooper pair."

The Cooper pairs travel through the crystal lattice, and the lattice, like a plucked violin string, vibrates at certain allowed frequencies. Just as a photon of light is created or absorbed when an electron changes its energy level in an atom, a particle of sound, called a phonon, is either created or absorbed when the lattice gains or loses energy. When one member of the Cooper pair collides with the lattice, it loses energy while the lattice gains it, creating a phonon. That phonon is immediately absorbed by the other electron of the pair. The second electron then loses its just-gained energy to the lattice, and the first one regains it. The pair of electrons continues to exchange phonons with each other, never losing energy as it travels through the lattice. Thus the electrons' collisions with the atoms become perfectly elastic, and they ride through the lattice together—without electrical resistance—on a microscopic sound wave.

For this remarkable theory, Bardeen, Cooper, and Schrieffer received in 1972 the second Nobel Prize awarded for superconductivity. Their theory not only explained the basis of superconductivity in metals, it also suggested where other scientists could look for new materials that would become superconducting at higher temperatures, perhaps high enough that liquid helium cooling would no longer be needed. For a while, there was steady progress, with a critical temperature reaching 23K just around the time of the BCS Nobel Prize. But then, it seemed, a limit was reached. Some scientists wrote articles explaining why they believed that superconductivity was unlikely at higher temperatures, and many laboratories even stopped working on it.

But still the subject held great fascination for many scientists. Even if they could not raise the temperature at which superconductivity would occur, they could learn more about the electrical properties of materials. Among those who persisted were Georg Bednorz and K. Alex Müller at the IBM Corporation's research laboratory in Zurich, Switzerland. They read about other scientists' research that had shown unexpectedly high superconducting temperatures in some materials. The temperatures were far below the record, but far above where theory had predicted them to be. Bednorz and Müller viewed those results as nature's hints of a path to materials that became superconducting at higher temperatures than had been previously seen. Following those hints, they began to study a class of ceramic materials known as perovskites, and in late 1986 they reported superconductivity in lanthanum barium copper oxide at 35K, a temperature 50 percent greater than the previous record.

The IBM Zurich discovery set off a flurry of activity in laboratories all over the world. Scientists worked night and day; some even slept on cots in their labs. Everyone wanted to be the first to discover superconductivity above the boiling temperature of liquid nitrogen (77K) or, with enormous good fortune, at room temperature (approximately 300K).

In 1987, Professor Paul C. W. Chu of the University of Houston announced that he and his associate M. K. Wu of the University of Alabama had discovered that yttrium barium copper oxide, with a chemical formula of $YBa_2Cu_3O_7$, was superconducting at a temperature of about 93K, or –180°C (about –290°F), 16°C above the boiling point of liquid nitrogen. Chu's announcement spread the excitement beyond the labs and onto the front pages of newspapers around the world.

That discovery, the papers reported, marked the beginning of a new era in superconductivity—and perhaps a new era in technology itself. No longer did all superconductors require cooling in expensive liquid helium; now liquid nitrogen could be used for some. Since liquid nitrogen is relatively inexpensive—it costs about as much as milk—engineers and business people who heard the announcement began to dream of the many new ways that these new superconducting materials might be used. They imagined superconductors in advanced computer chips, in high-speed trains that run with little friction on a magnetic cushion, and in wires for long-distance transmission of electric energy without the usual resistive losses.

Some of the dreams, such as the lossless transmission wires, required room-temperature superconductivity, but in the excitement, that temperature no longer seemed out of reach. In 1987 and 1988, new temperature records appeared almost weekly, as did new types of "high-temperature superconductors," as they came to be known. By the time the field began to settle down, the high-temperature record stood at a still frigid 125K (–148°C, or –234°F), and Bednorz and Müller had been awarded the 1987 Nobel Prize in physics, the third one given for work in superconductivity.

THE FUTURE OF SUPERCONDUCTIVITY

To many people who followed the superconductivity news in 1987 and 1988, the excitement has ended, but for scientists and engineers, the best still lies ahead. Will they need another theoretical breakthrough like BCS theory to understand the latest discoveries? Will other diligent scientists like Bednorz and Müller discover room-temperature superconductors, or have we reached the limiting temperature this time? Even if room-temperature superconductors are discovered, will we be able to use them in practical devices? For that matter, can we do anything useful with the 95–125K type of superconductor?

For nonscientists, the last question is the most important one. The new generation of superconductors have generated enormous interest, but they are not yet part of anything practical. Why? Part of the reason is that they, like other ceramics, are hard and brittle. Unlike metallic superconductors, they are very difficult to form into wires. Another concern is their so-called critical current, the amount of electricity they can carry before they lose their superconductivity. To increase that, we have to know their microstructure. For example, do they have crystal directions along which they are superconducting and others along which they are not?

Another important consideration is the relationship between their superconductivity and their magnetic properties. When a superconductor is exposed to a magnetic field in excess of a certain value called the "critical field," its resistance returns. That is very important in making superconducting magnets.

For use in computer chips and other electronic devices, engineers think not only of the possibility of tiny superconducting wires, but also of superconducting *thin films.* If they were to make a layer of superconducting material only a few atoms thick, or even a few hundred atoms thick, how would its properties differ from a thick, solid piece? Could they take advantage of some unusual properties, such as differences in conductivity in different directions, to make a useful device?

SUPERCONDUCTORS IN USE

Although the future of high-temperature superconductors is uncertain, low-temperature superconductivity already has its place in modern technology. Today's most powerful magnets are electromagnets made from superconducting wire. The materials from which that wire is made are the result of years of study of the properties and microstructure of superconducting metals and alloys. That study enabled people to make magnets from that wire in spite of limits such as critical currents and critical magnetic fields. Although liquid helium cooling is expensive, the power of these magnets makes them very valuable for certain applications and well worth the cost.

For example, as you read in the prologue, magnetic resonance imaging (MRI) gives physicians detailed pictures of their patients' internal organs in ways that X rays cannot. Furthermore, MRI carries less risk of long-term harm to the patients' health. Without powerful superconducting magnets, MRI would be far less effective, and its future would be limited.

Superconducting magnets are also important for particle accelerators, which scientists are using to unlock the secrets of the universe of subatomic particles. The Super Conducting Supercollider, which will be the most powerful particle accelerator ever built, will have hundreds of miles of superconducting wire with enormous amounts of liquid helium cooling. It will have so much wire, in fact, that some scientists think that its construction should be delayed until we know whether the new superconducting materials can be used. Others argue that people may never be able to make those materials into wires, and even if they were able, it would require too many years to wait.

Another use for superconducting magnets is in "maglev" (magnetic levitation) trains. These futuristic transportation systems are beginning to come into use in Germany and Japan. Maglev

This Japanese "maglev" (magnetic levitation) train set a world speed record of 517 km (321 miles) per hour. It used superconducting magnets to float above the tracks.

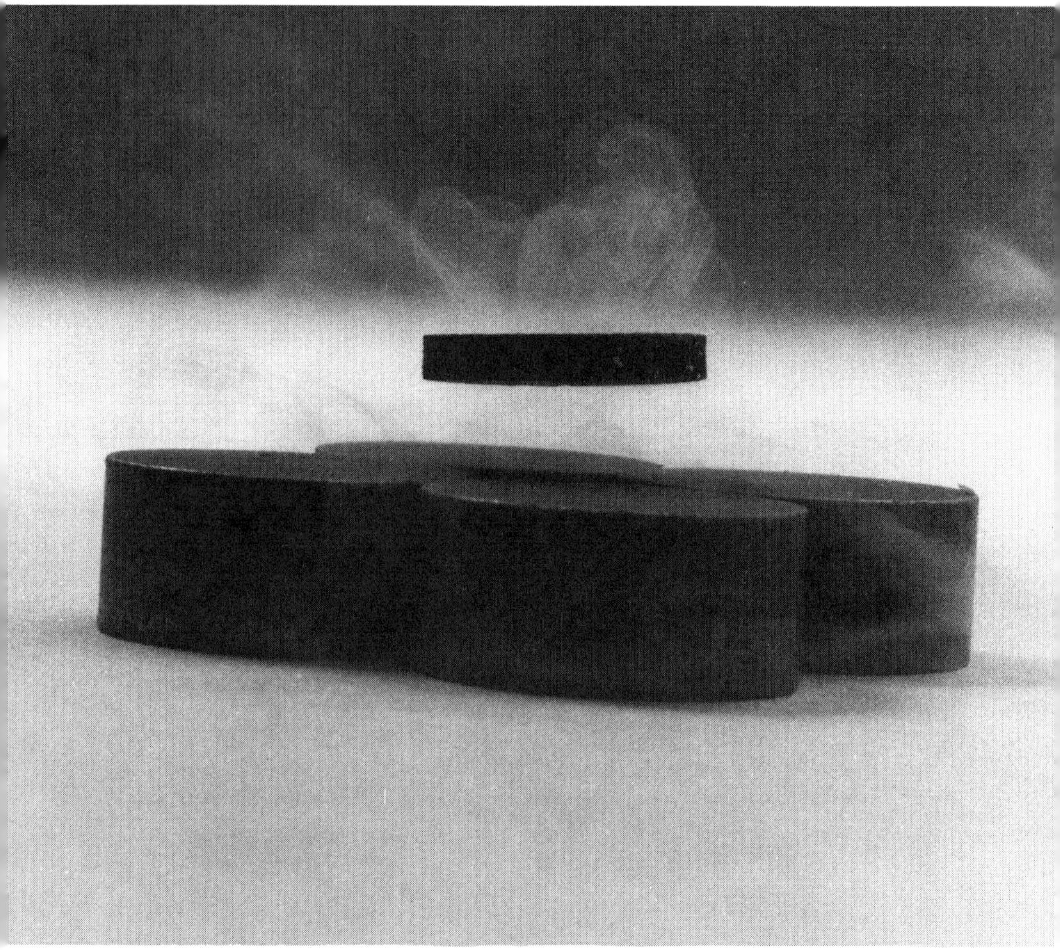

The Meissner effect. A superconducting pellet floats above an array of magnets. The pellets had been cooled to –321°F with liquid nitrogen. The magnets induced superconducting currents in the pellet, which created a counter-magnetic force to levitate the pellet.

trains use electromagnets to lift them toward the top of a T-shaped rail, and some train designs call for superconducting magnets. (Japanese Maglev trains use superconducting magnets, but German trains use ordinary copper wire.)

You may have seen pictures or demonstrations of a different kind of magnetic levitation of superconductors. The picture on page 70 shows a small piece of high-temperature superconductor floating in liquid nitrogen above a magnet. Like iron, the superconductor becomes a magnet when placed in a magnetic field. But unlike iron, it becomes magnetized in the opposite direction from the magnet creating the field, and is thus repelled. This behavior exists in some other materials, but only weakly. It is called "diamagnetism." A superconductor, however, is a "perfect diamagnet," developing a repelling field as strong as the one that magnetizes it.

This is called the Meissner effect, after the scientist who discovered it in the 1930s, and is considered the proof that a material is truly a superconductor and not just something with exceptionally low resistance. The Meissner effect has potential applications in magnetic levitation and as a magnetic shield (through which a magnetic field cannot penetrate), although no practical devices have been made that make use of it.

Chapter Five
Polymorphic, Polypresent Polymers

The last two chapters discussed materials with useful and unusual electrical properties. Now we turn to *polymers,* materials that have been developed because of their mechanical properties, such as density, strength, hardness, flexibility, stickiness, and smoothness, and the ability to be stretched or made into useful shapes. Just as semiconductors are useful because their electrical and optical behavior can be controlled so precisely, polymers are useful because their mechanical properties can be varied over enormous ranges.

In the 1967 movie *The Graduate,* there is a famous scene in which a successful executive calls aside a young man who has just graduated from college to offer him a one-word secret of success, "Plastics!" The executive's advice was good then, and it is good now, but we would change it to "Polymers!" because plastics are just one example of that much larger class of materials. You know polymers in many forms besides plastics: synthetic fabrics, nonstick coatings, adhesives, and different kinds of rubber, to name a few. Polymers also have many other uses, for example as lubricants or surfactants (molecules that attach themselves to and change the properties of the surface of a substance).

The word "polymer" comes from the Greek *poly,* meaning "many," and *meros,* meaning "part." The name describes the nature of the polymer molecule, which is, in fact, made up mainly of many repeating units, or *mers,* combined together. A single polymer molecule can have hundreds or thousands of repetitions of its basic unit. Because it is so long, scientists often use the term "macromolecule" (from the Greek *makros,* meaning "long") to describe it.

The properties of polymers, like those of all other materials, depend on their microstructure. In the case of polymers, much of the microstructure is in the makeup of the macromolecule itself. That microstructure is determined not only by the number of units and the atoms in those units, but also by the shape of the molecules. There can be substantial differences in microstructure—and thus in properties—between molecules that are part of the same polymer family.

THE CHEMISTRY OF CARBON

Nearly all of today's polymers are based on the chemistry of carbon. A few, such as silicones and polysilanes, are based on silicon, but we will focus

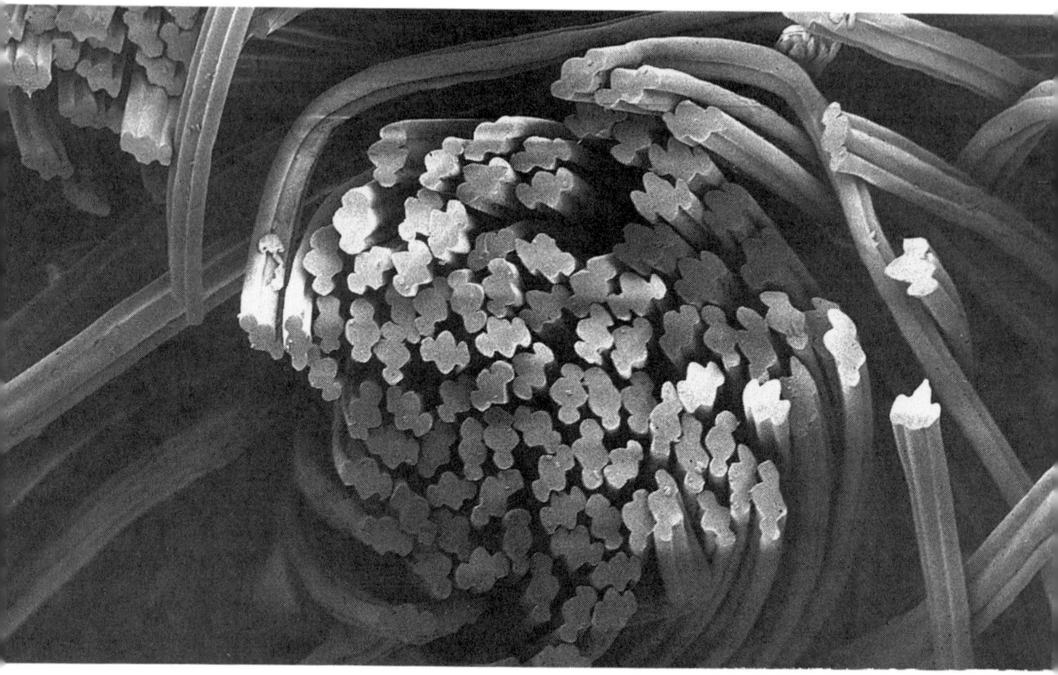

Above: four-channeled fibers of CoolMax, a synthetic fabric. The fibers transport perspiration away from the skin, making the fabric ideal for athletic wear.

Right: the Jarvik-7 artificial heart contains a special kind of polymer designed to inhibit blood clotting.

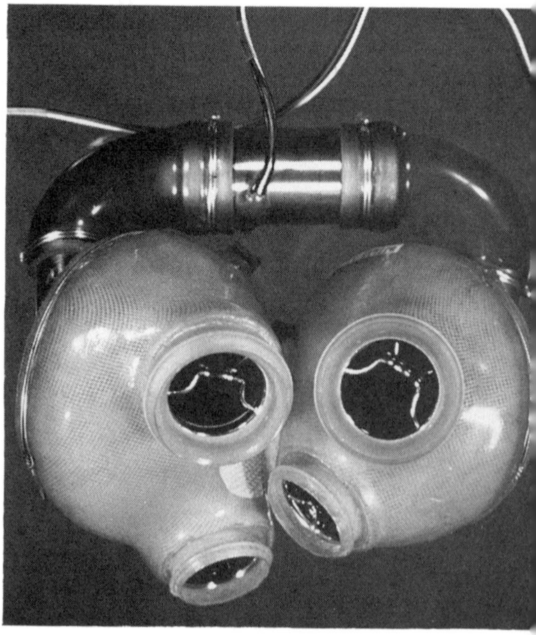

on carbon-based polymers. Each carbon atom has four valence electrons in a shell that is full when it contains eight electrons. As a result, it can form up to four covalent bonds in its compounds. Frequently, a carbon atom will combine with other carbon atoms in chains or rings, sometimes forming double or even triple covalent bonds with other carbon atoms and leaving fewer bonds available for combining with other atoms.

A simple introduction to the chemistry of carbon is the class of compounds called hydrocarbons, which contain only hydrogen and carbon. Let's start with the so-called saturated hydrocarbons, which have the maximum possible number of hydrogen atoms because all the carbon atoms form single covalent bonds. The simplest of these is methane, with the chemical formula CH_4. Then comes ethane, C_2H_6, propane, C_3H_8, and so on. Formulas like these tell how many of each type of atom a molecule has but do not show the bonds between the atoms. For that, chemists use structural formulas like these:

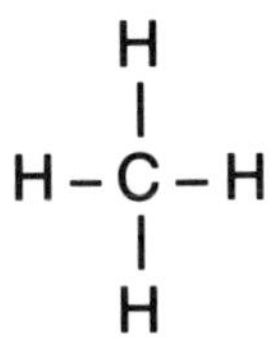

methane

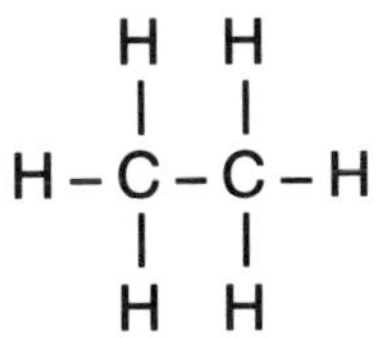

ethane

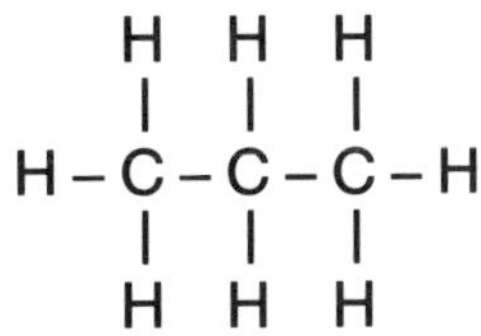

propane

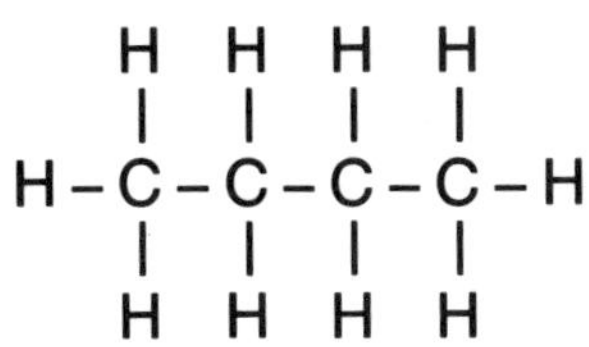

butane

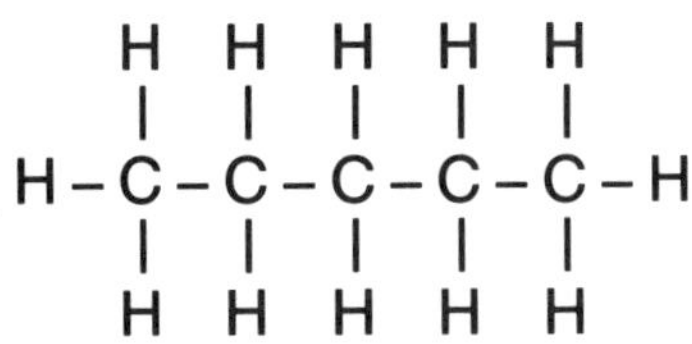

pentane

In the structural formulas, covalent bonds are shown by lines between atoms. Double covalent bonds are shown by a double line, and triple bonds are shown by a triple line. For example, here are the structural formulas for the unsaturated hydrocarbons ethylene, C_2H_4, and ethyne (or acetylene), C_2H_2:

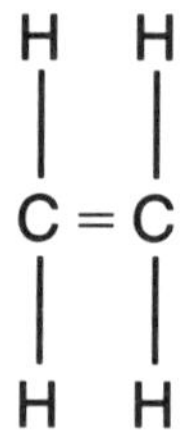

ethylene

H — C ≡ C — H

ethyne (or acetylene)

Often, the same chemical formula can result from two different structures, as for octane and isooctane, both C_8H_{18}, as shown here:

octane: 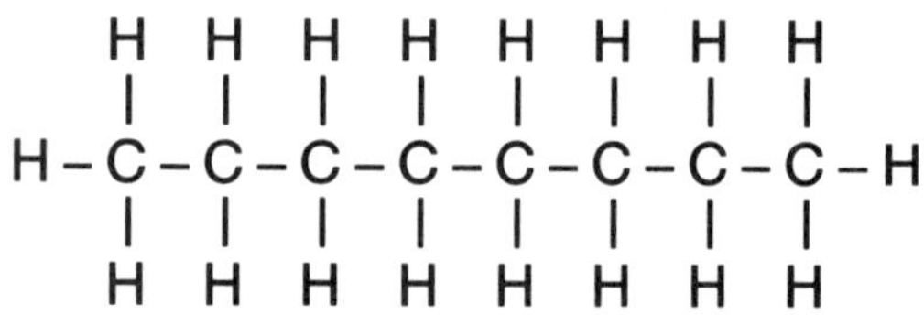

isooctane: 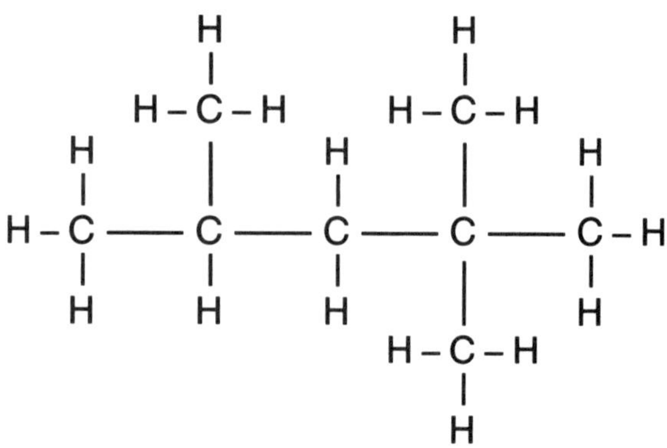

Notice that the carbons in the octane molecule make a straight line, while some of the carbons in the isooctane branch off the main chain. Thus, octane is said to be a linear molecule, while isooctane is called branched. In fact, you could describe isooctane as pentane that has had three of its hydrogens replaced by CH_3 (methyl) groups. Replacing one atom or group with another atom or group is a common way to make new carbon compounds. As you can imagine, chemists can use that approach to make larger and larger molecules—and to make an enormous variety of them.

When chemists do that, they are following a process that happens quite naturally in the development of living things. Thus, they often call the branch of chemistry dealing with the many varieties of carbon compounds "organic." Many compounds in naturally occurring organic materials, such as wood, rubber, and animal skin, are polymers. Even deoxyribonucleic acid—the famous DNA, which is the basis of life—is a polymer that contains two long coils of atoms linked by a series of units of four basic types. The order in which those units occur determines whether the life form is a fungus, a plant, an animal, or a human. If it is a human or animal, the DNA determines its sex, the color of its skin, hair, and eyes, and many other things about it. Unless you have an identical twin, your DNA is slightly different from everyone else's in the world.

With all that variety possible in the chemistry of carbon, it is no wonder that materials scientists and engineers have worked so hard to understand the properties of polymers and to create new ones. Let's examine some of the things they have learned and what they can do with their knowledge.

LINEAR POLYMERS

Suppose you were to take an atom of ethylene (C_2H_4) and partially break the double covalent bond between its two carbon atoms, leaving a single bond and two unbonded valence electrons, which we will call free electrons. Those free electrons can be viewed as partial bonds, ready to join with other free electrons, so the structural formula might look like this:

ethylene mer: 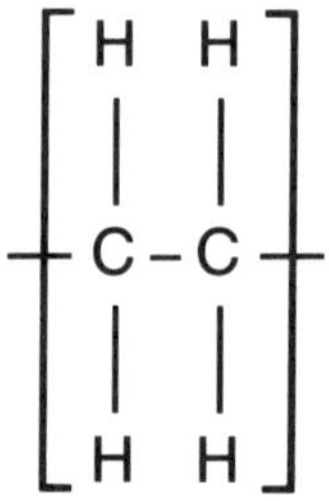

If two of these units were to come together, they would join, forming a chain of four carbon atoms attached to eight hydrogen atoms with free electrons at both ends. Other units could attach onto those ends, and the chain would continue to grow and grow. The very large chain eventually will stop growing when the free electrons at the ends join with something that has no free electrons of its own, but by then a macromolecule of polyethylene has been created.

Polyethylene is a very common plastic material. It is the simplest of the linear polymers, that is, those polymers with long chain molecules. Other well-known linear polymers are the vinyl family, in which one of the hydrogens of the ethylene mer (part) is replaced by a different atom or group. For example, the unit of polyvinyl chloride is simply this:

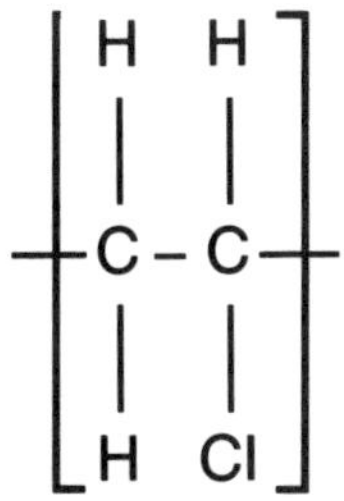

Polystyrene is another member of the vinyl family. You probably know it best as Styrofoam, made by blowing a gas through it as it forms. Styrofoam is used for disposable cups or plates and other insulating or packaging applications. Styrene, the basic unit of polystyrene, contains a group of six carbon atoms arranged in a closed loop with alternating single and double bonds. That arrangement is called a benzene ring, after the hydrocarbon benzene, C_6H_6, which has that structure. In polystyrene, the basic unit is formed by replacing one of the six hydrogen atoms of a benzene ring by a single bond attached to an ethylene mer, as the chlorine atom is in the polyvinyl chloride unit shown above.

One of the most famous polymers is the linear

macromolecule polytetrafluoroethylene, better known as Teflon. If you analyze its name, you may be able to figure out that all four of the hydrogen atoms of ethylene are replaced by fluorine atoms, so its unit looks like this:

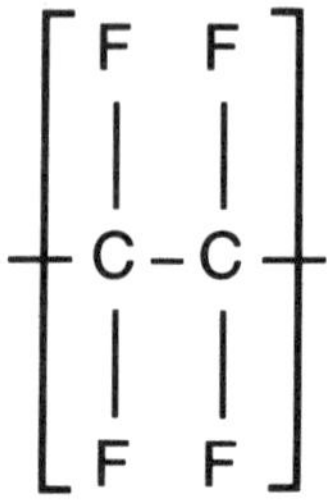

Another well-known class of linear polymers is the polyesters, which are useful in fiber form for clothing. Polyurethanes, also useful in foam form, are other linear polymers you may have heard of. Finally, rubbers are naturally occurring linear polymers. The differing mechanical properties of these substances relate to the lengths of the chains they form and the microstructure that results when those long chains interact with themselves and each other. (See the section titled "Microstructure" below.)

MAKING POLYMERS

Some polymers, like polyethylene, are made by addition of molecules with nothing left over. In order to make that happen, chemists have to do something to break a double bond and create the free electrons that start the polymerization process. Others are made by condensation, in which atoms or groups of atoms at opposite ends of a molecule combine to form a common small molecule, like water. The best-known example of this is the mer shown here in the box, based on a six-carbon chain:

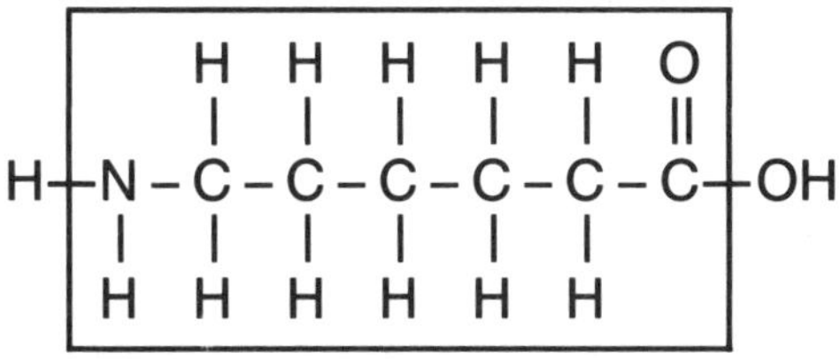

Under proper conditions, the hydrogen atom at the end of one molecule and the OH (hydroxyl) group at the end of another will combine to form a molecule of water, producing a chain of two mers, again with a hydrogen at one end and a hydroxyl at the other. The reaction will continue, producing larger chains of mers and more water molecules.

The polymer product produced in this condensation reaction is nylon, a strong, shiny fiber that revolutionized the way people dressed. Before the invention of nylon in 1938, expensive silk stockings and shirts were the symbol of wealth and elegance; afterward, the use of the new fabric nylon meant that almost everyone could afford clothes that looked smooth and shiny. Nylon became so popular as a stocking fabric that women called their stockings "nylons." But clothing was not all nylon was good for. Nylon cords were so strong and durable that they were used in automobile tires, among other things.

MICROSTRUCTURE

Many of the unusual properties of polymeric materials can be understood by thinking about the unusual microstructure that can result from their large molecules. Linear polymer molecules look and behave much like a chain of beads. If the chain is long enough, it can organize itself into patterns, similar to crystal lattices but without the three-dimensional connections. Likewise, many chains together can lead to regular arrangements of the mers.

Depending on the shape of the mers and the tightness with which they are bound together, the chains can be either relatively stiff or flexible, relatively straight or kinky. Rubber's stretchability can be understood from the shape of its molecules, which are very kinky chains. When you pull on a piece of rubber, you straighten the kinks of its molecules!

Natural rubber is too stretchable—it can be described as gooey—to be used in tires or other things where some stiffness or resiliency is desirable. You may have heard the story that Charles Goodyear discovered, quite by accident, that adding some sulfur to rubber gives it resiliency without loss of its flexibility. He called his product "vulcanized" rubber and went on to make a fortune in the tire business.

Today we understand that the strength comes from chemical links that the sulfur forms between different parts of a linear molecule. Think of a chain of a thousand beads that has arranged itself into a snakelike S shape so that the 57th bead is in contact with the 288th bead on a different arm of the S, and many other pairs of beads are in similar contact. By pulling on the ends of the chain, it would be easy to break those connections. So the chain could easily be stretched to many times its length, and it would be slow to return to its old shape. When it does return, bead number 57 may now be in contact with bead number 240.

But if you were to put springs between some of those beads on different parts of the S, you would make the connections permanent and enable the chain to return to its old shape more quickly. That's the way vulcanizing works. By adding chemical connections between mers that happen to be near each other, sulfur gives natural rubber resiliency.

Chemists call that process *cross-linking.* Cross-linking adds a three-dimensional structure that makes some polymers very hard. It is also the process that makes superadhesives work; they form a joint that is often stronger than the materials they are connecting by a chemical reaction that creates a tangle of cross-links.

Chapter Six
Metals and Alloys: Shiny, Strong, and Still Super

Although smelting, alloying, and forming metals are among our oldest technologies, the science and engineering of metallurgy are relatively new. People are continuing to learn more about metals and to find new ways to use them. Metals, which caught ancient people's attention because they were shiny, strong, and easily shaped, are still super in the high-technology world of the late twentieth century.

Strength and shapeability are still major reasons people use metals, but other properties are also important. Some metals, like aluminum and magnesium, are useful because they are relatively light. Others, like lead, tungsten, and uranium, are useful because they are heavy. Tungsten is also useful because it gets hot enough to glow brightly without melting, while lead-containing alloys like solder and Wood's metal (used in sprinkler systems) are useful because they melt at relatively low temperatures.

There are metals, like copper, silver, and gold, that people use when they need a good conductor of heat or electricity. But when people make a heating element, they need a not-so-good electrical conductor that can get very hot without melting, like the alloy called nichrome. If they need something powerfully magnetic, there are alloys that are far more magnetic then iron. If they need something nonmagnetic, there are many metals and alloys to fill the bill.

As that list of useful properties tells you, metals and alloys, like semiconductors and polymers, are useful because of their versatility and because people have developed the ability to control their properties. And what is the basis of that ability to control the properties of metals and alloys? By now you can probably guess the answer: changing their microstructure. The object of the science of metallurgy is to understand the microstructure of metals and alloys and to learn how to change it. Metallurgical engineering is using that understanding and that ability to change metallic microstructures in the creation of useful materials.

SINGLE-PHASE AND MULTIPHASE METALLIC MATERIALS

To begin to understand the properties of metals and alloys, we start with a reminder that metals are crystalline materials. We must also remember that although people usually think of crystals as perfect, many of the important properties of real crystalline materials are determined by imperfections. That is certainly true of metals and alloys. In fact, *on the microscopic scale, a metal or alloy may not be a single material at all.*

To understand what we mean by that surprising statement, think about an industrial diamond, which is nearly pure carbon, most of it in the diamond crystal structure, but with some inclusions of graphite. The graphite inclusions ruin its value as a gemstone but don't affect its hardness or its usefulness as a cutting tool, so we can still call it a diamond. If it were half graphite and half diamond, with tiny regions of the two structures intermixed, we could no longer consider it a diamond with imperfections, nor could we consider it graphite. We would simply have to call it a two-phase piece of carbon, where the word *phase* is used to refer to a state of matter with a certain arrangement (or randomness) of its molecules.

Many metallic materials are like that, with regions of two or more different materials intermixed. The regions may have the same chemical composition with different crystal structures or they may have entirely different chemical compositions. Such metals and alloys are called *multiphase.*

PROPERTIES OF SINGLE-PHASE METALS AND ALLOYS

Single-phase metallic materials can be either a nearly pure metal or one of two types of alloy: (1) a solid solution, in which the atoms of two or more different metals are mixed at random positions in a crystal lattice; or (2) an intermetallic compound, in which the atoms of two or more metals have a definite arrangement on the crystal lattice (Figure 8). (There are also two noncrystalline types of single-phase alloys called amorphous alloys or metallic glasses and quasicrystals, but these are not common today.)

Bronze, which was the first alloy discovered by humans, is a solid solution of tin in copper with about 5 percent of the atoms being tin. Brass, a solid solution of zinc in copper, was discovered around the beginning of the Iron Age. (If it had been discovered before iron smelting, there would probably have been a Brass Age.) It can

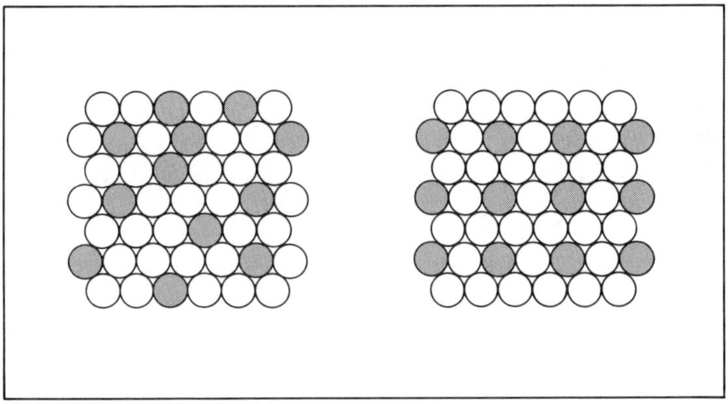

Figure 8. Alloys can be either solid solutions or intermetallic compounds. In a solid solution (left), the atoms of the different types of metal are located on lattice sites in a more or less random order. In an intermetallic compound (right), the arrangement of atoms follows a definite pattern.

contain up to 40 percent zinc. Copper and zinc also form an intermetallic compound CuZn, called beta-brass. Sterling silver is a solid solution of 7.5 percent copper and 92.5 percent silver.

A solid solution can be thought of as a crystal of mainly one type of atom with a fraction of the lattice sites occupied by the other atoms. If the other atoms are similar to the main atoms in size and chemical properties, they can be added in large amounts without upsetting the overall regularity of the crystal. However, just as you can dissolve only a certain amount of salt in water, so you can add only a certain amount of one metal to another and form a single-phase solid solution. Adding more produces a multiphase material. For example, if you add a very little bit more than 7.5 percent copper to silver, you no longer get a single-phase alloy. Instead, you get mostly regions of the sterling silver phase mixed with regions that are primarily copper with about 8 percent silver in them—a two-phase alloy.

People make alloys to create materials with superior properties for certain applications. For example, adding copper to silver produces the alloy sterling silver, a material much more wear-resistant than pure silver and just as shiny. (Sterling silver was originally developed for use as coins.) The modern discovery that a small amount of copper hardens silver is similar to the Stone Age discovery that a small amount of tin hardens copper. Stone Age people found the new material, bronze, to be superior to copper alone because it was hard enough for use as tools. The alloy's discovery thereby revolutionized civilization and launched the Bronze Age.

The same hardening effect occurs for many alloys besides bronze and sterling silver. How does alloying make a metal tougher? The answer, as usual, comes down to microstructure. The crystal structure of the metal is bound to contain some dislocations. When you apply a force to the crystal, atoms can slide easily along a dislocation in such a way that the dislocation moves through the crystal. In addition, applying the force tends to create more dislocations. The moving dislocations make a metal soft and easy to wear away.

As long as all the atoms in a crystal are the same, that picture of easily moving dislocations holds true. But as soon as you add atoms of a different size, the situation changes. Trying to put a larger or smaller atom into a crystal lattice produces a distortion. Likewise, if a crystal lattice already has a distortion, it is easier to fit in an odd-size atom. Thus if a moving dislocation, which is a distortion in the crystal, encounters a different-size atom, the dislocation tends to get stuck—or pinned—in place by that atom. Adding the alloying metal in solid solution has therefore made it more difficult to move the dislocation, and has thus made the material more resistant to wear.

That's not the only effect of alloying, of course. Metals are excellent conductors of heat and electricity due to the free movement of conduction electrons. The planes in the crystal lattice act like channels for the electrons to move in. An atom of a different type distorts the channel, so a moving electron is more likely to collide with it. As a result, electrons take longer to carry their charge and energy through the material and therefore solid-solution alloys tend to be poorer conductors of heat and electricity than pure metals.

If an alloy is an intermetallic compound, it tends to be especially hard (Figure 8). Unlike in a pure metal or a solid solution, the sliding of one plane of atoms in an intermetallic compound with respect to another plane of atoms makes a significant change in the crystal. In beta-brass, for example, copper atoms in one plane are chemically bonded to zinc atoms in the adjacent planes. To move a plane with respect to its neighboring plane, you would first have to break all the copper-zinc bonds and move through an intermediate state where the atoms are adjacent to their own kind rather than the opposite kind, and then reestablish the copper-zinc bonds two atoms away from their original position. To keep one plane moving along the other, you would constantly have to break those bonds as soon as they form again.

FATIGUE

When you speak of the properties of a material, you usually refer to a behavior that can be measured very quickly: its hardness, its electrical or thermal conductivity, its ability to stretch or flex without permanent changes in shape, its brittle-

Fatigue failure of a railroad rail. The fatigue started at the top left. Initially it grew slowly. The larger rings show how the fatigue process sped up over time.

ness or its ability to be drawn into wires, its melting temperature, the way it interacts with light, its magnetic behavior, and so on. But there are other properties that are important because they measure changes in the behavior of a material or an object made from that material over a relatively long time.

For example, an object may flex back and forth millions of times over a period of years. Then suddenly, it fractures instead of flexing. When scientists examine the break under a microscope, they find that it developed microscopic cracks that joined together into one large one where the object came apart. Or, for example, a metal tube in a boiler may withstand the high pressures and temperatures of superheated steam for years, seemingly without changing. But suddenly it cracks, and steam rushes out explosively. When engineers examine the broken pipe, they discover that it had become thin in the area that failed. In both of those cases the failure was sudden—or "catastrophic," in the vocabulary of scientists and engineers—but the changes that led to it had been occurring unnoticed for years.

The phenomenon of fracturing instead of flexing is called *fatigue* because the material appears to get tired. The phenomenon of movement of material under steady pressure, which can lead to thinning, is called *creep*. You can demonstrate fatigue for yourself with a wire coat hanger. Bend the long, straight portion of the hanger into a sharp angle, then bend it back until it makes a sharp angle in the other direction. Repeat this back-and-forth bending for a few minutes and you will feel the hanger begin to flex more easily with less springiness. Soon after that, it will crack at the place where it was bending. While you are flexing, stop every now and then to feel the temperature at the point of bending. You will notice that the wire is getting hot. The heat comes from the friction of grains moving against one another. As they move, tiny cracks begin to develop between them. Eventually the cracks join together to produce the break in the hanger. (You might think that the increase in temperature is responsible for the fatigue, but it is not. You could prove that by flexing the hanger as much but doing it more slowly so it stays cool.)

Engineers guard against fatigue somewhat by choosing materials with microstructures that are less likely to crack. But their main defense against it is in the detail of the shape near the points of bending. By proper design, they can keep the bending angles to a minimum and so can place less demand on the microstructure. You can understand the importance of that design process by flexing a coat hanger a little bit instead of a lot. You will find it impossible to cause it to fatigue because you will not bend it enough to start the microscopic cracking.

STEEL, AN IMPORTANT MULTIPHASE ALLOY

Without question, steel is the most widely used, versatile, and important metal product in the history of human civilization. Its major component is iron, the second most common metal in the earth's crust, and one which is easily extracted from its ore by smelting. (Aluminum, the most common metal, requires enormous amounts of electrical energy. Furthermore, most aluminum occurs in clays which are not suitable as ore.) The term *steel* actually includes a large family of iron-based alloys. Besides iron, all steels contain about one-half of 1 percent carbon (and no more than 1.2 percent in any case). By adding varying amounts of other elements (up to 5 percent), steelmakers produce a huge variety of products. Cast irons, which contain 2.5 to 4 percent carbon, are generally considered members of a different—but related—alloy family.

The combination of iron's abundance and the strength and ease of manufacture of cast iron and steel led to the development of Iron Age civilization and made possible the Industrial Revolution. Today, plain carbon steel is still manufactured in great quantity and is used as sheet metal for automobile bodies and home appliances, among other things.

Unlike easily shaped sheet metal sheets, certain other steels are extremely hard and tough enough to serve as bulldozer blades and high-strength gears. Some steels are designed to be used in electrical transformers and function as the core of an electromagnet that reverses its magnetization many times each second, while steels used in wristwatches must be completely nonmagnetic. While iron rusts easily, some stainless steels are able to withstand strong acids without corroding.

Steel is a complex multiphase material. Its strength comes from grains of the ceramic material iron carbide (Fe_3C), also called cementite. (Ceramics, as you will read in the next chapter, are usually very hard, but very brittle materials.) The microstructure of steel begins with pearlite, which is a two-phase iron-cementite material containing about 0.8 percent carbon. Pearlite's microstructure consists of alternate layers of iron and cementite, with the iron phase being the greater part of the material. Because the brittle cementite is surrounded by iron, pearlite retains the ability of iron to be shaped while gaining considerable toughness from the hard ceramic phase.

Steel generally has a different carbon content than pearlite, so it forms a more complex microstructure. If the overall carbon content of the steel is less than that of pearlite, its microstructure consists of grains of pearlite mixed with grains of iron. If the overall carbon content of the pearlite is less than that of the steel, the microstructure is grains of pearlite separated by thin bands of cementite. In either case, it is a two-phase microstructure in which one of the phases (the pearlite) is itself a two-phase structure. As you see, simply by changing the carbon content, it is possible to produce a wide range of microstructures and properties of steel.

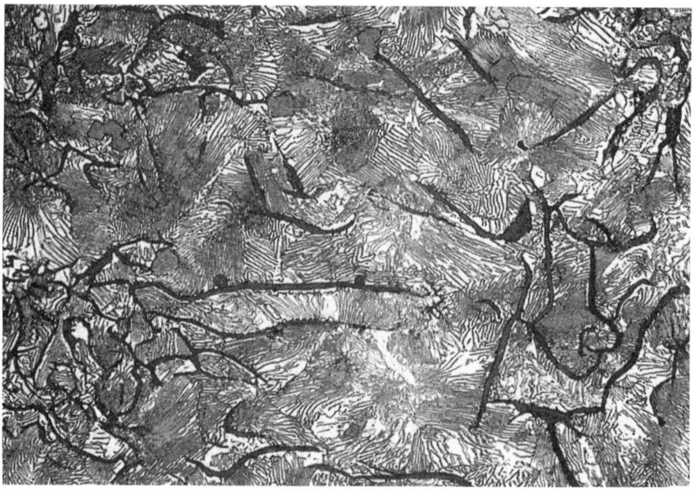

Compare these magnified photos. Top shows microstructure of pearlite, a component of steel. Bottom shows gray cast iron, consisting of a pearlite matrix with graphite flakes.

METALLURGICAL PROCESSES

Metallurgy begins with smelting, the process of extracting a metal from its ore. In the ore, the metal is combined with a nonmetal, most commonly oxygen or sulfur. Heating the ore to a high temperature in the presence of some other material causes a chemical reaction that causes the nonmetal atoms to combine with other atoms and leaves the metal free. Smelting is not always that simple. For example, iron ore is not pure iron oxide; it contains many impurities, especially *silicates* (compounds of metals, silicon, and oxygen). In smelting the iron, these impurities are removed by the addition of limestone (calcium carbonate).

Electrolysis

Electrolysis, the process of running an electric current through a compound to reverse the chemical reaction that formed it, is another way of extracting a metal from its ore. Because it requires an enormous amount of electrical energy, electrolysis is used only when smelting will not work. The only metal produced in large quantities in this way is aluminum.

Alloying and Quenching

Once the metal is smelted from its ore, it may be mixed in molten form with other metals, forming alloys. These may be, when cooled, solid solutions or intermetallic compounds, and they may be single- or multiple-phase materials. Making a solid solution is not always a simple matter of mixing molten metals. The amount of one molten metal that can be dissolved in another to make a solution depends on the temperature. Thus once a molten solution of metals is made, it is often advantageous to cool it very quickly, in a process called *quenching,* which preserves the high-temperature composition because the atoms do not have a chance to rearrange themselves before the alloy solidifies.

Not only the composition but also other elements of the microstructure of an alloy depend on how slowly or rapidly it is cooled from its liquid state. If it is quenched, different regions begin to crystallize separately and the atoms don't have much time to adjust their positions to match the developing crystal structure, so grains tend to be small and crystal defects common. If it is cooled slowly, larger, more-perfect grains are formed. If you want to make nearly perfect large single crystals, you must freeze the material very slowly.

Cold- and Hot-Working

One of the most important discoveries of the Bronze Age was the strengthening of metals by beating them, especially when they are hot. We now know that such changes are due to the motion of crystal defects. Heating a metal gives each atom, on the average, more energy. That added energy makes it more likely that an atom will occasionally leave its place on the crystal lattice, thereby creating more point defects and making

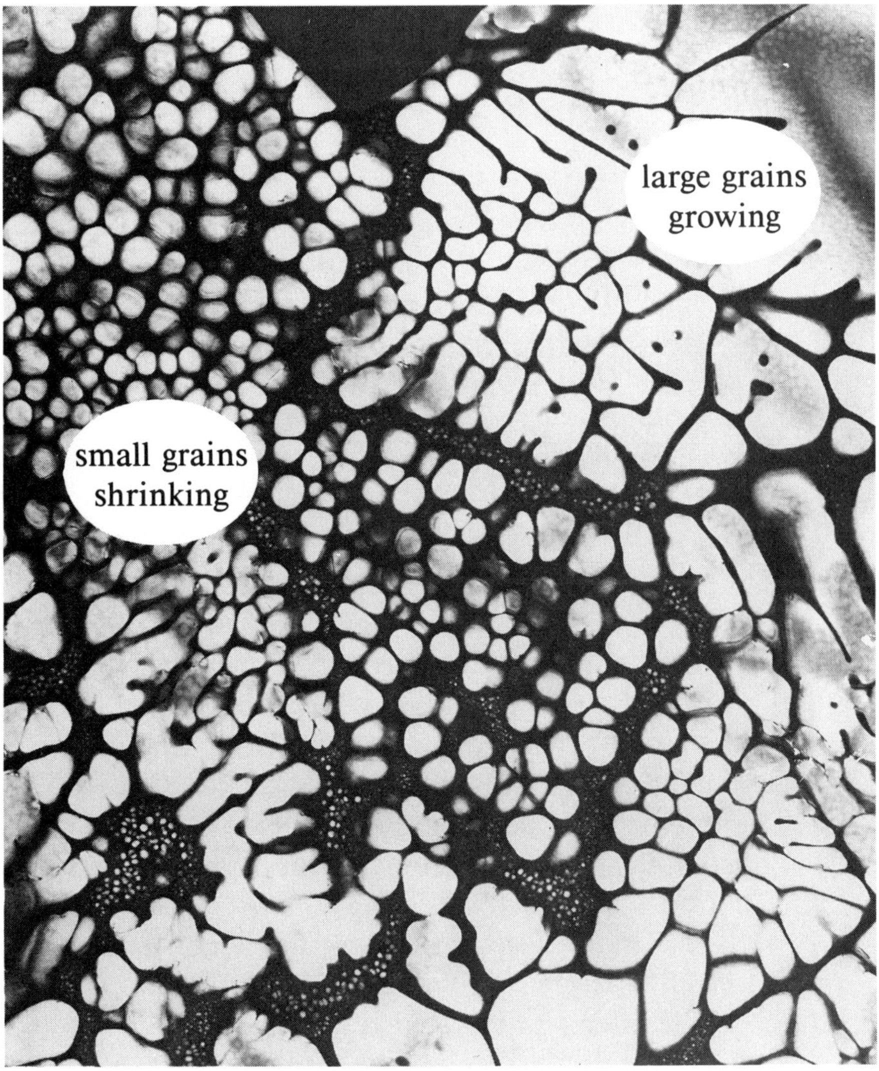

A transmission-electron micrograph (TEM) of a nickel-aluminum-silicon alloy, showing large grains growing and small ones dissolving.

them more likely to move about. Applying stresses to the metal by beating or rolling it makes its crystal defects move. Sometimes, stress can create new defects, such as dislocations. The combination of heating and beating can exaggerate these effects. The presence of dislocations can make a material softer, but the presence of enough dislocations to create tangles can ultimately lead to greater toughness. The processes of applying mechanical stresses, with and without the presence of heat, are called *hot-working* and *cold-working* respectively.

Annealing

Another process that can lead to significant changes in microstructure, and thus in the properties, of metallic materials is *annealing*. Annealing is heating the material to an elevated temperature, short of the melting point, and letting it remain at that temperature for an extended period of time. Annealing permits the process of diffusion, the gradual movement from place to place of atoms and crystal defects, to create significant changes in microstructure. One frequent result of annealing is grain growth.

The effects of annealing, quenching, and mechanical working on the properties of metals have been known for many centuries, but only recently have we come to understand their effects on microstructure. With that understanding, we continue to create new metallic materials with made-to-order electrical, magnetic, and mechanical properties.

Chapter Seven
Ceramics and Glass

Human civilization has a two-million-year history, the first 99.7 percent of which is what we have called the Stone Age. Putting the history of civilization on a time scale comparable to your own life, the Stone Age ended about two weeks ago. So although you may think of it as ancient history, you should also realize that the Stone Age shaped human behavior in many ways that continue today.

In fact, there is something deceptive about saying that the Stone Age ended when the Bronze Age began. Bronze made great tools and weapons in many ways, but stone and other ceramic items still had many significant advantages, and they were made from the most common natural materials. So although bronze revolutionized civilization, stone and ceramic items continued to be very important. Bronze and other alloys did not replace ceramics; rather, they supplemented them. In fact, although the new materials got most of the attention, people continued to work with the old ones, finding new and better ways to make and use them.

THE CHEMISTRY AND MICROSTRUCTURE OF CERAMICS AND GLASSES

Ceramics and glasses are different forms of the same chemical substances. Ceramics are crystalline while glass is *amorphous.* "Amorphous," which means without form, describes an arrangement of atoms and molecules in an irregular pattern, as if they are in a liquid, although the substance is solid (that is, its atoms or molecules do not move past one another). Ceramic substances are chemical compounds of metals (or semimetals like silicon) and nonmetals, bound together by powerful chemical bonds which are partially ionic and partially covalent. The bonds get their strength from the fact that they usually involve two or more electrons from each atom or group of atoms.

Aluminum oxide (two aluminum atoms bound to three oxygens) and silicon dioxide (one silicon atom bound to two oxygens) are both ceramic materials. In fact, most ceramics are oxides or silicates (compounds containing the silicate group—a common chemical unit made of one silicon atom plus three oxygen atoms). This should not be surprising, since oxygen and silicon are the two most common atoms in the earth's crust (47 percent and

28 percent respectively, by weight). Aluminum is also common in ceramics because it is the third most prevalent atom in the earth's crust (although, by weight, there is less aluminum than iron). The rocks of the earth are natural ceramics (or glasses, in the case of volcanic rocks like obsidian).

The strong bonds are responsible for ceramic materials' most useful properties. For example, ceramics can withstand high pressures because the bonds make it difficult to change the distances between atoms. Also ceramics can withstand very high temperatures without melting because it takes more thermal energy before the atoms vibrate enough to disrupt the crystal structure. Finally ceramics (and glasses) are highly resistant to corrosion. That is because corrosion is a chemical reaction in which metallic atoms or ions gain electrons, but in ceramics, the metals' atoms have already gained electrons as part of those bonds.

Another characteristic property of ceramics is their brittleness. They are so hard that they will break along a crack rather than stretch or bend when pulled or twisted. The breaks follow grain boundaries or other crystal defects.

APPLICATIONS OF CERAMICS

The hardness, heat resistance, and corrosion resistance of ceramics clearly make them attractive materials. But what about the brittleness? For us, brittleness can be a disadvantage, but for early humans, who had only the ceramics that nature provided—rocks and stones—brittleness was an asset. It enabled them to break off pieces of the stones to make sharp-edged tools.

Today, most ceramics in use are not natural stone. Rather, like all other modern materials, they are made in ways that control their microstructure to enhance desirable properties and minimize undesirable ones. Here are a few of the important applications of ceramics.

Refractory Materials

Materials that can withstand very high temperatures are called *refractory.* Refractory materials are most commonly used as linings for furnaces, so they must not only have high melting points and low thermal conductivity but also maintain their strength at high temperatures.

The tiles that protect the United States' space shuttles from the heat of reentry are made of a specially developed refractory ceramic material. Previously, when space vehicles were not reusable, heat protection was provided by a material that flaked off as it got hot, carrying the heat away. The space shuttle tiles provide permanent protection.

Abrasives

Grinding and cutting are common industrial processes. Ceramics, because of their hardness, are commonly used on grinding wheels and cutting tools. Natural and synthetic (manufactured) gem-

The space shuttle is covered with heat-resistant ceramic tiles.

stones and such man-made ceramics as the carbides of boron, silicon, titanium, and tungsten are important abrasive materials.

Insulators

Ceramics are excellent electrical insulators because the bonds between the atoms tie up the conduction electrons quite effectively. Power lines are usually protected by insulators made of electrical porcelain, a ceramic not very different from that used for plumbing fixtures. Not only does that ceramic have exceptionally high electrical resistance; it can also withstand a high voltage without "breakdown" (a sudden surge of electricity through a material in much the way lightning flashes through the air when the voltage between two nearby clouds is high enough).

Piezoelectric Materials

Some ceramics exhibit a property called piezoelectricity, which gets its name from the Greek word *piezein,* meaning "push." A *piezoelectric* material produces an electric current or voltage when the pressure on it changes. In a piezoelectric crystal, the positive and negative ions in the substance tend to be on opposite sides of each lattice site. When you change the pressure, the oppositely charged ions move a little closer together or farther apart, so there is a pulse of electric current.

Piezoelectric crystals have many uses as pressure sensors. One of the most familiar is in a phonograph cartridge. They are also used as vibrators, especially to produce ultrasound (sound at a frequency higher than what humans can hear). When the voltage across a piezoelectric crystal changes, the spacing between the two types of ions changes in response, producing motion in anything in contact with the crystal.

Superconductors

The new high-temperature superconductors are a ceramic material. They are typically an oxide of three metals (such as yttrium, barium, and copper) in the ratio of one atom to two atoms to three atoms of the metals to seven atoms of oxygen. Their crystal structure is similar to that of the natural mineral perovskite (calcium titanate, $CaTiO_3$), and they are therefore called perovskites. Apparently they get their superconductivity because a small fraction of the oxygen atoms in their crystal structure is missing, with the result that regions of the crystal are left with excesses or deficiencies of electric charge. Forming ceramic superconductors into wires is difficult because of their brittleness.

Magnets

When most people think of magnets, they think of iron. But experts in magnetics know that many of the most useful magnets are ceramics. In fact, the famous lodestone, the first known naturally magnetic material, is a ceramic—iron oxide. In electronics, a ceramic magnet can often have an advantage over a metallic magnet because it is an electrical insulator.

Construction and Household Materials

Ceramic products are used in so many valuable ways that even if we were to ignore their "high-technology" applications, they would remain important. For example, the versatile ceramics are used not only as construction and household materials such as bricks, tiles, tableware, and plumbing fixtures, but also to create a variety of beautiful pieces of art.

SILICA AND SILICATES

When silicon and oxygen combine, it is in the form of the chemical compound silica or in the chemical unit called a silicate ion. Silica is silicon dioxide (SiO_2), and the silicate ion is a chemical combination of a silicon atom and three oxygens plus two electrons to complete its outer shell. Silicate ions combine with metal ions to produce substances called *silicates.* Silica and silicates are the most important substances in the ceramics and glass industries.

Like carbon, silicon is very versatile in forming compounds because its atoms have four electrons in their outermost shell. Because of that, the microstructure of silicates is in some ways similar to that of polymers. For example, silicate ions can form long chains like this:

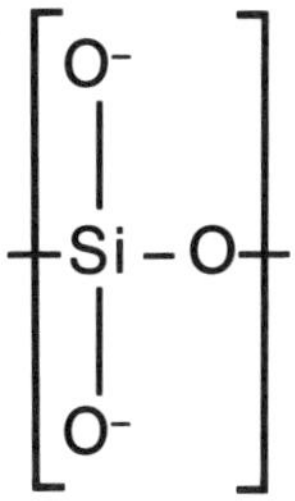

silicate ion

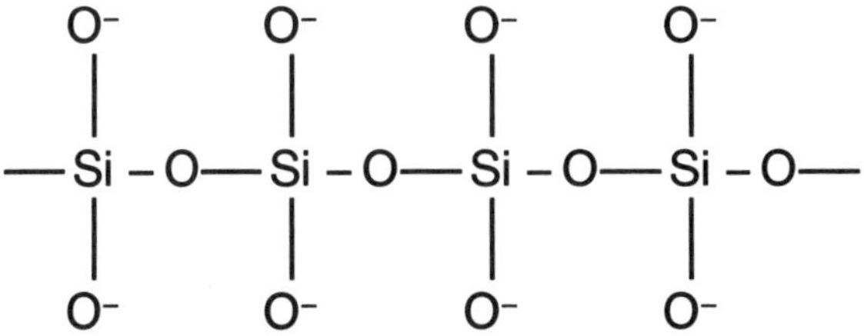

silicate chain

You will notice some similarities between the silicate chain and the polyethylene mer shown on page 78, but there are significant differences as well. The main similarity is the capability to form very long chains. The main difference is that the oxygen atoms at the top and bottom of the silicate chain are shown with a little minus sign, indicating that they each carry an extra electron that is ready to enter into a chemical bond with a metal or semimetal. That is because the silicate unit is an ion. (Sometimes chemists cause those electrons to become part of a covalent bond with an organic unit instead of a bond with a metal. That forms the well-known silicate-based polymers called silicones.)

Another important difference is in the way the molecules are shaped in three dimensions. In hydrocarbon polymer molecules, the hydrogen atoms do not favor any particular arrangement, so the molecules behave much like strings. In silicates, each silicon atom sits in the center of a tetrahedron, a four-sided solid like a pyramid with a triangular base, with oxygen atoms on each corner. The oxygen atoms in the body of the chain are called bridging oxygens because they serve as connections between two tetrahedra. Thus silicate chains are more like interlocking tetrahedral beads.

Furthermore, in silica and silicate materials, these chains are linked by metal or silicon atoms forming crystalline or amorphous arrangements of interconnected tetrahedra. The wonderful variety of those arrangements gives silicates their versatility.

For example, in the minerals mica and talc, as well as in natural clays, three corners of each tetrahedron are shared with others, forming sheet-like structures. That explains why mica breaks easily into thin flakes and talc can be slippery.

Silica occurs naturally as sand, quartz, and other minerals in three-dimensional crystalline arrangements of linked tetrahedra (Figure 9). It is the most common example of the kind of minerals known as network silicates. The mineral feldspar is another network silicate. It is the pink phase of the well-known stone material granite and has the chemical composition potassium aluminum silicate ($KAlSi_3O_8$). The aluminum ion replaces a silicon atom in the silica crystal structure, but since aluminum has only three electrons

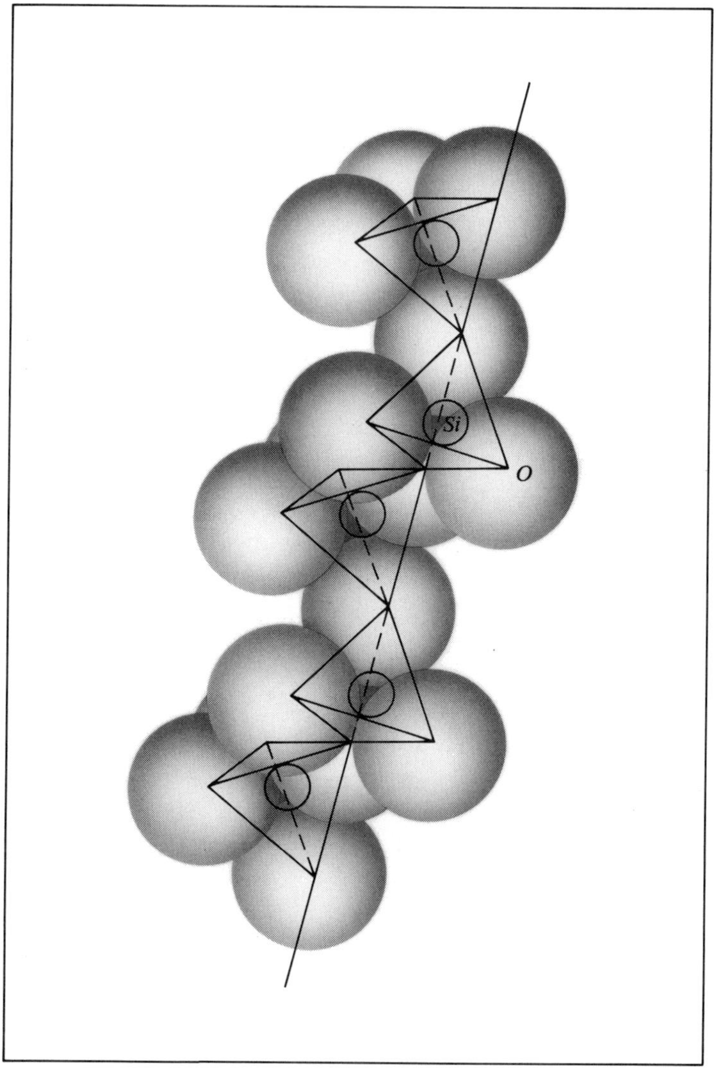

Figure 9. Silicates have a basic unit of three oxygen atoms and one silicon atom arranged as a tetrahedron. Silicones have chains of such tetrahedra. Sheetlike arrangements of silicate units comprise such minerals as mica and talc and the crystallites in clay.

in its outer shell, the potassium, with its single outer-shell electron, is needed to balance the charges. The network crystal structure does not have a definite place for the potassium atom. But fortunately, since the network is quite open, there are plenty of *interstitial* locations where it can fit.

Glassmaking begins with network silicates, usually silica in the form of sand. When silica is heated in the presence of "network modifiers," which are usually oxides like calcium oxide from heated calcium carbonate (lime) or sodium oxide from heated sodium carbonate (soda ash), the extra oxygen atoms, being nonbridging, lower the melting point. The network modifiers also add metal ions, which disrupt the network by reducing cross-linking between silicate chains, further lowering the melting point. The larger number of nonbridging oxygens and the reduced cross-linking leads to an amorphous state when the material solidifies again.

PROCESSING OF CERAMICS AND GLASS

The only ceramic process known to the ancients was adding water to clay and heating it in a fire. The water enabled the clay to stick together and to be easily shaped. The heating removed the water and made the material hard as stone. Other processes to make glass and ceramic products require heat, pressure, and controlled addition of chemicals. These processes are easy to understand in principle. Yet carrying them out has become precise and sophisticated as scientists gain improved understanding of the relationships between the properties and microstructure of these materials.

For example, ceramic engineers have developed many processes to make solid objects from powdered ceramics by pressing the powder in a mold, then cementing the granules together by heating. That process is called sintering. In some cases, the heating forms little glassy regions that join the powder particles together. In other cases, the joining takes place by actual transfer of atoms from neighboring particles which gain enough thermal energy to move across the regions of contact.

Because ceramic particles are so hard, applying pressure to them does not change their shape, and the particles in powdered ceramics are already so tiny that they will not break when squeezed together. Thus any ceramic product made from powder is bound to have empty spaces between the particles. This can have some advantages; for example, lower weight without much loss of strength. However this can also create a porous product or one with too little strength for the job.

How can a ceramic engineer make a solid product from powder with just the right amount of space between the particles? If a powder is made up of particles of approximately the same size, a large fraction is bound to be empty space. You can understand that by putting a collection of marbles in a container until you can get no more in. Now if you try pouring water in (make sure the container won't leak!), you will find that you have a lot of empty space between the marbles. That will be true even if you use sand instead of marbles, although it will be less dramatic. But if you use a mixture of sand and marbles, you will fill much more of the volume of the container. Where there is a marble, it fills up its space completely, so it is better there than the sand would be. Where there is no marble, at least you can fit in sand. So the trick is proper choice of the mix of particle sizes.

The ceramic engineer uses the same solution, a mixture of both coarse and fine particles, so the fine particles can fill the spaces between the coarse ones. That's neat and simple, but it requires a lot of scientific work to get detailed knowledge of the sizes of the particles, and it requires creative engineering to get the desired mixture of particle sizes.

Glass products are formed by heating them until they reach a temperature at which they soften, then applying forces to form them as desired. Unlike a crystalline material which changes all of a sudden from solid to liquid, glass behaves differently, mainly because it is amorphous. When glass softens, it does not change in the same way as a melting solid, whose molecules go from an orderly arrangement to a disorderly one. Both above and below the softening temperature, the molecules are about as orderly as those of a liquid. Nearby atoms may be arranged in a somewhat orderly pattern, but there is certainly nothing like a crystal lattice that can be orderly over millions of atoms.

When the glass is soft, people work on it by stretching it, cutting it, blowing air into it, drawing it into a fine fiber, adding chemicals to change its color, and so forth. They will also anneal it to let thermal energy produce changes in their product's microstructure. In some cases, they will even permit it to form crystalline regions so it will cool as a true solid. When that happens, the product is called a ceramic glass.

HIGH-TECHNOLOGY GLASS

Glassmaking and glass-forming techniques have been developed over centuries. In that sense, it is an ancient art. But with the many "high-tech" applications of today, it is also a very modern technology. Aside from the main applications of glass in ordinary windows, containers, tableware, and optical equipment, here is a sampling of the exciting uses of that remarkable material called glass.

Tempered Glass

Your mental image of glass is probably that it is a material that breaks easily, like your neighbor's kitchen window that shattered when you hit an unexpectedly long home run. Yet glass can be made tough enough to be used in high-strength applications such as the walls of modern metal-and-glass skyscrapers. One way to strengthen glass is a process called tempering.

In that process, the object is formed from a piece of glass that is cooled quickly on the surface while the inside remains hot. The outside then becomes like a solid bottle in the shape of the object, while the glass inside is still hot and soft. As the inside glass cools, it contracts and pulls inward on the bottle. When the cooling is complete, the surface of the object is permanently under a stress that is trying to compress it while the glass in the center of the object experiences a stretching force.

Now imagine that a projectile like a baseball —or perhaps even a bullet—hits the glass. The impact will cause the surface to experience a stretching force while the inside experiences a compressive one. The internal forces of the glass will counteract those of the impact, and the projectile will bounce off without damage.

Glass Fibers for Insulation

People today are very conscious about the amount of energy they use. Insulating homes and buildings to minimize heat loss or to retain cool, conditioned air has become very important. The most widely used material for home and building insulation is a woolly material made of long, thin glass fibers.

High-Tech Cookware

The microwave oven has become a very common item in the high-technology home. Microwave cooking requires utensils that do not absorb microwaves, so metal is out and glass is in! In the microwave oven, the food gets hot but the pot does not.

But most people who use microwave ovens still use their stove top and ordinary oven at other times. They would prefer not to have to buy different sets of pots and pans for microwave and ordinary cooking. Glass manufacturers think they have solved that problem. Now there are glass pots that do not melt at temperatures that turn metal pots to soup. There is even a television commercial that shows a glass pot with a metal pot softening and bending inside it.

Optical Fibers for Medicine, Computers, and Communication

"Fiberoptics" refers to the use of carefully crafted fibers of glass that act as pipelines to carry light around corners and into areas that could never before be illuminated. The inside of the fiber is one type of glass and the outside is a different type. That design makes it possible to send a beam of light into one end of the fiber and have it bounce off the inside of the glass, follow twists and turns, and come out the other end.

The first major use of fiberoptics was in many different kinds of medical tools to examine areas inside the body that previously could be seen only during surgery. That breakthrough has saved many lives and eliminated many painful operations.

Today, the most exciting use of fiberoptics is in computation and communications. In both fields, information of one kind or another, from telephone voice signals to computer data, is encoded as a series of on-off pulses of light. Specially designed *optical fibers,* with carefully produced variations in the chemical composition of the glass, can now transmit those pulses 35 miles (56 kilometers) at a time, eliminating the need for 80 percent of the costly "repeaters" that were formerly needed in communications cables. An optical fiber can carry four thousand times as many calls as a copper cable of the same size, and its weight is so much less that a ship can now carry and lay a 3,000-mile (4,800-km) undersea cable in a single trip. Formerly, several trips were needed. These improvements are a major reason for the great increase in the use of telephone communications between people and computers and the decrease in the cost of long-distance calls.

New products of ceramic materials and glass enable us to travel into space, to heal our bodies, to cook our food, to conduct electricity, to compute, to communicate, and to deal with information in many ways that could only be imagined a few years ago.

Chapter Eight
Superstuff, Present and Future

As the twentieth century comes to an end, people naturally wonder what the next century will bring. In any field of technology, the future is hard to predict in detail, but its general shape can be seen in the most advanced products and in the laboratories of the present. In *materials science* and *materials engineering,* the future will also be shaped by an array of remarkable new tools and techniques. So first we will look at the tools and then at the most advanced materials in the world today.

THE TOOLS OF SUPERSTUFF

The tools of modern materials science and engineering enable people to understand and control the microstructure of a material almost atom by atom. To remind you how small an atom is, a drop of water contains about 1,000,000,000,000,000,000,000 molecules (that's one thousand million million million, or one sextillion, if you like to name big numbers that way). Or if you like to think of atoms lined up side-by-side, a line of two hundred and fifty million atoms would be about an inch (2.54 centimeters) long. If you weigh 100 pounds (45 kilograms), your body contains about 2,000,000,000,000,000,000,000,000,000 (two octillion) atoms.

You have probably also read that atoms and molecules are too small for people to see. That statement is no longer true! Although people cannot see individual atoms with the naked eye or ordinary microscopes, the newest tools of materials science can sense atoms in many different ways and produce images that show individual atoms or the shape of single molecules. And if you can sense single atoms, you can begin to consider manipulating them one at a time.

Not long ago, these amazing tools existed in only the most advanced laboratories as one-of-a-kind instruments built by the scientists themselves. Today they are available for purchase from manufacturers. With the tools of today and the tools to come, materials scientists and engineers know that practically nothing about microstructure is too small to detect and that only the laws of nature and their imagination limit the properties of materials they can make.

These instruments fall into two categories according to their function. Characterization tools are those which measure and detect things about materials. They help scientists and engineers learn their characteristics. Fabrication equipment is what people use to make—fabricate—materials and devices.

CHARACTERIZATION TOOLS

Because individual atoms are several thousand times smaller than the wavelength of visible light, ordinary optical microscopes are of only limited value to materials scientists. However, when large-scale features of the microstructure, such as grain size and structure, are important, optical microscopes are still the tool of choice. But with advanced materials, it is often necessary to know about the microstructure on a much more detailed level.

The first step in the investigation is the determination of the basic crystal structure. Mathematical analysis of the geometry of solids proves that there are only seven possible basic shapes for the building blocks of crystals, all of which have sides that are either squares, rectangles, parallelograms, triangles, or hexagons. People have found natural substances with all seven of these basic crystal structures, but no others, just as the analysis predicts. The main technique used to determine crystal structure is *X-ray crystallography.* When X rays reflect from a crystal to a piece of photographic film, they produce a pattern of dots that reveals the basic crystal structure and the spacing between the atoms on the crystal lattice.

X-ray crystallography combined with chemical analysis tells a lot about a material, but it does not provide much information about the imperfections in the microstructure. Chemical analysis can indicate that impurities are present, and blurred X-ray diffraction pictures can show that the material is not a perfect crystal. But if you are looking for dislocations, you need to use a microscope. A normal optical microscope won't work, because dislocations are smaller than light waves. What's needed is a microscope that uses something else. That something else is electrons, and that microscope is called an electron microscope.

The branch of physics known as *quantum mechanics* tells us that all particles have wavelike properties and all waves have particlelike properties. Two examples are light waves (made up of particles called photons) and the nucleus of an atom (surrounded by a cloud of electrons). The wavelength of an electron is related to its energy. The higher the energy, the shorter the wavelength. In a technique called *electron microscopy,* electron microscopes use electrons with high energy to form images of structures as small as crystal planes and dislocations.

An electron microscope can also be used to perform microchemical analysis. When its electrons strike a region of the material, that region gives off X rays. The wavelengths and intensity of the X rays tell scientists the chemical composition of that region. By similarly analyzing other regions of the sample, the scientists can determine how its chemical composition varies from place to place. This is especially important for materials whose properties are sensitive to small amounts of impurities.

Just as both optical and electron microscopes have their advantages, so do they have their disadvantages. For example, both microscopes produce images of either surfaces or very thin samples. But suppose a certain type of integrated circuit fails repeatedly in use and the designer suspects that a defect beneath the surface may be the cause of the problem. Is there some kind of device that can look inside? The answer is yes, acoustic microscopes, which use ultrahigh-frequency sound waves, can. Acoustic microscopes are built and sold for problems of this sort.

In fact, many physical phenomena can be used to create images. Imaging technology is probably the most rapidly changing area that affects materials science and engineering. In one notable example of this technology, imaging is based on a strange quantum mechanical effect known as tunneling. The images are produced by a device called the scanning tunneling microscope (STM). Other microscopes based on the STM but with capabilities for characterizing and imaging other properties of materials on the atomic scale (for instance, magnetic fields) are currently finding use in laboratories all over the world. No one knows where these new imaging capabilities will end or what they will reveal.

FABRICATION TOOLS

Characterization of new materials and devices is only half the story. Making them is the other half. Here are a few examples of present and future

Above: a scanning-tunneling microscope (STM). Right: a silicon surface revealed by the STM, showing locations of individual atoms.

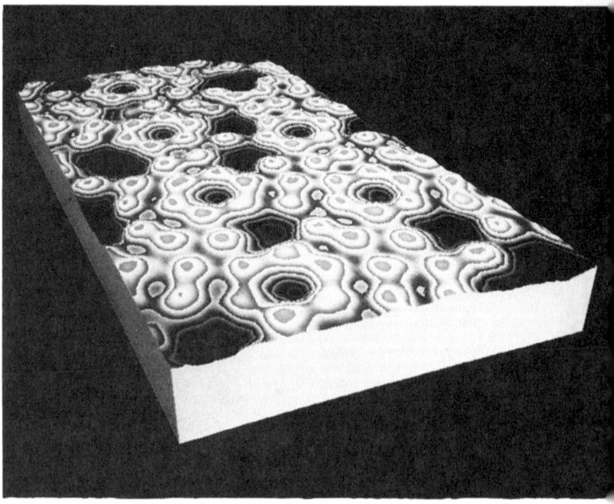

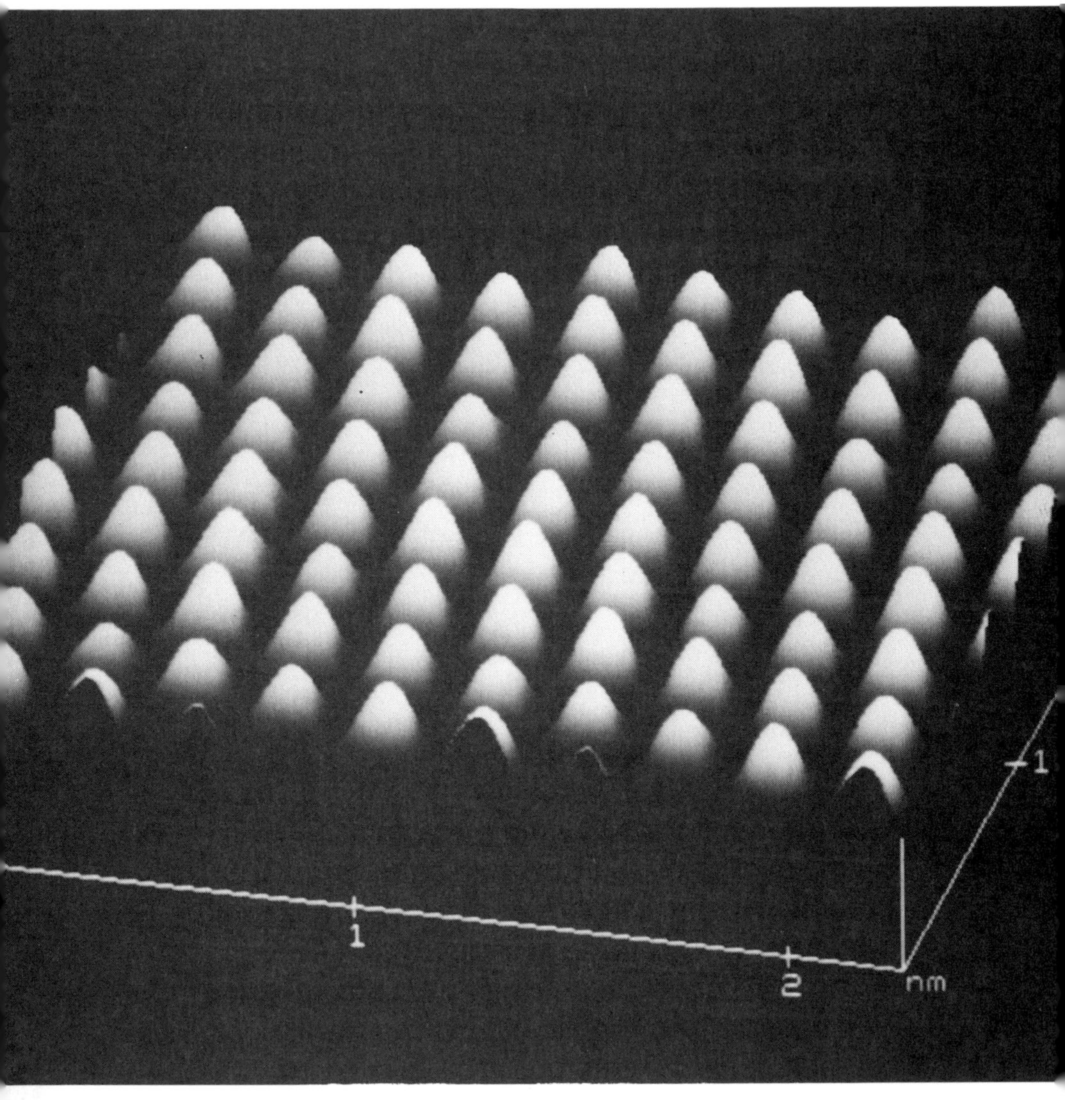

Carbon atoms on the surface of graphite, revealed by a scanning-tunneling microscope.

superstuff fabrication tools and the wonders they can or may produce.

In Chapter Three, you read about *integrated circuits (ICs).* ICs are an example of devices made by *microfabrication,* which includes a whole class of processes to make very small devices. Current microfabrication techniques are based mainly on precise photography, so the transistors on an IC can be made no smaller than the wavelength of light. To make smaller parts on future ICs, X rays and electron beams will be used.

Microfabrication is becoming important for industrial products other than integrated circuits. The most important product of that sort is a tiny electromagnet used to "read" and "write" information on magnetic recording disks in computer systems. Tiny mechanical parts for unbelievably small motors are also beginning to come off production lines. Some scientists are fabricating tiny devices that are small enough to take advantage of properties predicted by quantum mechanics. These "quantum wells" are expected to lead to breakthroughs in electronics and photonics.

Because the properties of certain materials can be greatly changed by small, controlled amounts of impurities or lattice defects, the technique of ion implantation was developed. In ion implantation, the impurity atoms to be added are exposed to a large electric field which removes electrons and turns them into ions. They are then accelerated by a large electric field and sent crashing into the material that is to be modified, creating crystal defects and adding impurities where they are desired. Ion implantation is used as part of the microfabrication process for semiconductors and in creating very specialized and very sensitive detectors.

SUPERCOMPUTERS

"Supercomputers?" you ask. "Computers are wonderful machines, but they can't be used to make superstuff, so why include them?" You are correct that computers can't make materials, but you must not forget that people can now design materials with made-to-order properties. Designing materials often includes a great deal of theoretical analysis and computation. Supercomputers, combined with imaginative human minds and exciting new characterization and fabrication techniques, have been important and will continue to be important tools of the Age of Superstuff.

ADVANCED MATERIALS OF TODAY AND TOMORROW

Let's peer into the future a little bit by looking at a few areas of materials science in which research is most promising. Most of these areas have already produced significant breakthroughs, but their promise is even greater than their present.

Composites

Composite materials are, in one sense, very old. Ancient people discovered that adding straw to mud made better bricks because the straw prevented the brittle ceramic from cracking all the way through. Modern ceramic materials, like ancient bricks, are brittle. So to take advantage of ceramics' strength, heat resistance, and light weight, materials scientists and engineers are developing ways to add strong yet flexible materials, like microscopic graphite fibers, to ceramics. The materials they produce in this way are called *composites.* They are also working on composites based on plastics and metals. They classify composites by the main material they are made of, so they use the terms "ceramic matrix composites," "polymer matrix composites," and "metal matrix composites."

Ceramic matrix composites are already being used to make lightweight yet powerful aircraft engines. If there is a breakthrough that reduces their cost, you may someday drive a car with a composite engine.

Plastic objects are generally light and easy to manufacture, but they usually are not strong or resilient enough to be used where mechanical stresses are high, such as in the bodies of automobiles or airplanes. Polymer matrix composites designed to provide strength are likely to take the place of plastics in future manufactured products. The ultralight *Voyager* airplane that flew around the world in 1986 without refueling could not have done so without plastic composites.

Steel, a multiphase alloy, is in effect a metal-matrix composite with little regions of ceramic added to give it strength. Materials scientists are

The ultralight Voyager aircraft was flown nonstop around the world by Dick Rutan and Jeana Yeager. It is made of a polymer matrix composite, which combines the lightness of plastic with the strength of more durable materials.

working toward future breakthroughs in all areas of composites, and they are certain to change our everyday lives in many significant ways. One everyday area in which composites have already made their mark is sports. Because of composite materials technology, you can now buy much lighter, stronger, more flexible tennis racquets and golf clubs than you could only a few years ago.

Rapidly Solidified Materials

When a liquid freezes, the degree of crystalline perfection in the solid that forms depends on how rapidly and how uniformly it cools. If it cools extremely rapidly, the atoms may not have sufficient time to assume their normal crystal pattern. The resulting solid might have many microscopic grains, it might have regions of another phase that would normally not exist at the lower temperature, or it might be amorphous. The microstructures often lead to unusual and often very useful electrical, mechanical, or magnetic properties.

Many materials scientists and engineers are experimenting with various techniques to solidify materials extremely rapidly, and they often make surprising discoveries. For example, in 1985, a group of scientists from the Johns Hopkins University and the United States Bureau of Standards were working on the rapid solidification of an alloy of aluminum and manganese. When they examined their product with X-ray crystallography, they were startled to see an X-ray diffraction pattern that was different from any previously seen. The atoms had apparently arranged themselves in a pattern shaped like a twenty-sided solid called an icosahedron, which was supposedly impossible for a crystal because a repeating arrangement of icosahedrons always leaves gaps of different shapes between them. Since there is no pattern that repeats again and again, you can't call the arrangement crystalline; but since it is made up of blocks of the same shape, it is almost crystalline. Scientists call it a quasicrystal.

The discovery of quasicrystals has led many laboratories to study rapid solidification. That work has been very fruitful from both the scientific and engineering point of view. Whether or not we ever have quasicrystalline superstuff, rapidly solidified solids will undoubtedly lead to valuable discoveries for a long time to come.

Thin Films, Interfaces, and Multilayers

Strange as it may seem, even a perfect crystal is imperfect in a very important way. If a crystal were perfect, it would look exactly the same from every lattice site. But that can be true only if the crystal goes on forever. Real crystals, however, end at their surfaces. The lattice sites at the surface have neighbors only in one direction and not the other. That makes the surface a very dramatic crystal defect, and causes the properties of the surface of a material to be very different from the properties of the material as a whole. Because of that, a large branch of materials science and engineering involves surface studies.

A *thin film* of material has a large fraction of

its atoms near the surface. The thinner it is, the more unusual its properties are because the more like a surface it becomes. Besides behaving in ways that are different and potentially useful for making a device, thin films contain so little volume that people can consider making them out of very scarce substances. The cost of a thin film is usually more in the equipment used to make it than in the material that it contains.

Thin films are only the starting point of many research studies. Since the properties of a material are largely determined by the neighbors of its atoms, a surface would be expected to have different properties if it is in air, water, or a vacuum, or if it is in contact with another surface. An *interface* is a region where two different surfaces come together, and the properties of interfaces are often even more unusual than the properties of surfaces. Thus, scientists who study surfaces often study interfaces as well.

Many future breakthroughs are likely to come from surface and interface studies, but some scientists aren't stopping there. Because materials fabrication tools now enable them to control the composition and microstructure of a material very precisely, they are letting their imaginations consider what would happen if a material were made up of thin films of different types stacked up layer by layer. The properties could change dramatically if the thickness of the layers were to change. These multilayer structures are likely to be the source of many future breakthrough materials.

Thin films can be made in several ways. One of the most effective ways is sputtering. In a vacuum chamber, the materials engineer places a target of one or more different substances and bombards it with high-energy ions. Those ions knock atoms loose from the target, and those atoms collect on a base material called a substrate. Sputtering enables materials scientists and engineers to control the properties of thin-film materials by varying the composition of the target, the energy of the bombarding ions, and other factors in the manufacturing process. Thin films made in this way are becoming especially important as the information-storing coatings on magnetic disks and as very thin friction-reducing and corrosion-protective coatings for high-technology products.

A more common technique to produce thin metallic coatings on a large scale is vapor deposition. In that case, the material that is to form the coating is vaporized by heating. The vapor then collects as a thin film on the substrate that is to be coated. Vapor deposition is much less expensive than sputtering, but sputtering gives more-precise control of the properties of interest.

Some materials are made by adding atoms to the surface of an already existing crystal. The added atoms assume the crystal structure of the base material, although they might not ordinarily do so. The added molecules may be either in a liquid form or a vapor form. This is called epitaxial growth or epitaxy. Liquid-phase epitaxy is used to make magnetic bubble memories for computers.

The newest fabrication tool for epitaxial growth is the molecular beam epitaxy (MBE) machine. It enables materials scientists and engineers to grow one material on another at a carefully controlled rate. Using MBE, scientists are creating materials with layers a few atoms (or molecules) thick of one substance in a stack with similar layers of another substance. Theoretical analysis of these multilayer materials suggests that they will have unusual or previously unattainable properties of great use in new electrical, magnetic, or mechanical devices.

Manufacturing in a Weightless Environment

Gravity is often a problem in manufacturing. An object inside an orbiting satellite is still influenced by the Earth's gravity, but it seems to be weightless because both it and its surroundings are falling at exactly the same rate. (They don't fall to Earth because they are also moving parallel to the surface, which curves away beneath them as fast as they are falling.) Thus, an artificial satellite is a place where people do experiments on the effects of weightlessness.

A number of space flights have included experiments in making high-purity materials with fewer crystal defects. As people establish permanent orbiting laboratories, materials science and engineering will be high-priority areas of study. Breakthroughs in space-made superstuff are not only likely, but expected.

SUPERSTUFF AND YOU

What lies ahead for superstuff? No one can make specific predictions, but everyone in materials science and engineering is sure that there will be more excitement of all kinds: breakthroughs, discoveries, surprises, inventions, rewards for the people who work in the field, and benefits for everyone. The study and application of materials offer a wealth of areas to explore, and that exploration will involve many people with many different talents: scientists, engineers, computer scientists, technicians, writers, and business people, to name a few. You may work in the field, but even if you don't, your life will be affected by it. You live in a world of "high technology," and technology has always been based on the materials at hand.

We humans have come a long way since we first discovered what we can do with the materials of nature, and we've come a long way since we first discovered how to make our own. Now that we can see atoms and move them around, where will we go from here? No one knows the answer, but it will surely be lots of fun to find out as we enter the Age of Superstuff.

Glossary

Absolute zero. A temperature of approximately −273° Celsius, or −459° Fahrenheit—so cold that it is impossible to reach, although it can be approached. Scientists usually measure low temperatures in Kelvin units, which are equivalent to Celsius degrees above absolute zero.

Alloy. A material that results from mixing two or more metals. It has properties that are often significantly different from the metals that go into it.

Amorphous. A type of microstructure in which the atoms of the material are not arranged in any regular pattern. Glass is the best-known amorphous material.

Annealing. A process used in the processing of materials, especially alloys and glass, in which the material is heated to just below its melting point and held there for a length of time during which the microstructure can change significantly through the movement of atoms within the material.

BCS theory. A theory proposed by physicists John Bardeen, Leon N. Cooper, and John Robert Schrieffer that explains the phenomenon of superconductivity in certain materials.

Bronze Age. The period of human history immediately following the Stone Age. During the Stone Age the discovery of the alloy bronze revolutionized civilization and ushered in the Bronze Age by making possible many new or improved tools, implements, or weapons.

Ceramics. A class of materials that are chemical compounds of metals or semimetals and nonmetals. Ceramics are generally light, strong, and resistant to corrosion and heat. They are also usually quite brittle.

Cold-working. The process of applying mechanical stresses to a material, usually a metal or alloy, without heating it. Cold-working can harden a material by changing its microstructure, mainly by moving its dislocations and creating new ones. See **dislocation.**

Composites. A class of man-made multiphase materials in which the main phase (the matrix) is either ceramic, metal, or polymer, and the second phase is often a fiber such as graphite or a tiny particle. The additional material is added to lessen the effect of an undesirable property, such as brittleness, in the matrix material.

Conductivity. The ability of a material to permit an electric current (electrical conductivity) or heat (thermal conductivity) to pass through it.

Covalent bond. A link formed between two atoms by their sharing of electrons.

Cross-linking. The formation of chemical bonds between two widely separated units of a polymer molecule, producing a three-dimensional microstructure and major changes in properties.

Crystal. An orderly, repeating arrangement of the atoms or molecules of a material that often results in its ability to be cut into pieces with smooth faces and sharp edges or to form naturally into such pieces.

Defect, or **imperfection.** A region of a crystal in which the orderly arrangement of its atoms or molecules is disrupted.

Dislocation. A crystal defect in which the atomic planes are slightly offset from each other.

Doping. The controlled introduction of an impurity into a material. Doping is especially important for creating semiconductor devices with desirable properties.

Electrolysis. A process in which the passage of an electric current through a liquid reverses the chemical reaction that caused a compound to form. Aluminum is produced by electrolysis of molten aluminum oxide.

Electronics. A branch of technology based on the ability to control the flow of signals or energy by controlling the motion of electrons. Semiconductor technology has revolutionized human civilization through the development of electronic devices with many new capabilities.

Electron microscopy. A technique for producing images of very small regions using the wave properties of electrons. Electron microscopy has made it possible to see the microstructure of materials and even individual atoms.

Energy band. According to quantum mechanics, a range of energy levels for electrons in solids, most important of which are the valence band and the conduction band.

Energy level. According to quantum mechanics, one of the energy values that an electron in an atom can have. Only certain values of energy are possible.

Fatigue. A phenomenon in which a piece of metal fractures or cracks, apparently suddenly, after being stressed a large number of times repeatedly in the same way. Fatigue occurs because the microstructure gradually changes due to the stresses, until eventually the material can no longer withstand them.

Glass. An amorphous material generally made from silicon dioxide with the addition of a few other substances. Scientists sometimes use the term "glass" to describe any amorphous material.

Grain. One of many small crystalline regions of a material that is made up of many such regions.

Hot-working. The process of applying mechanical stresses to a material, usually a metal or alloy, while heating it. Hot-working changes the properties and microstructure of a material by causing the motion of its defects or by creating new defects.

Imperfection. See **defect.**

Impurity. A crystal defect consisting of an atom of a type that is not part of the normal chemical makeup of the crystal, such as an atom of aluminum in a crystal of silicon.

Inclusion. A small region of a crystal that is different chemically or in crystal structure from the material surrounding it. Inclusions of graphite diminish the gem-quality value of a diamond.

Industrial Revolution. The recent period of human history in which the development of steam power and advances in steelmaking led to great changes in industry and society.

Insulator. A material that has a low conductivity for heat or electricity.

Integrated circuit (IC). A device that contains a large number of transistors or other electronic components on a single piece of semiconductor material. It is produced by microfabrication techniques. See **microfabrication.**

Interface. The region where two different surfaces meet.

Interstitial. A crystal defect consisting of an atom in between normal crystal lattice sites.

Ionic bond. A bond between two atoms in which one gives up one or more electrons to the other, the transfer forming a negative ion and a positive ion that are held together by an electrical attraction.

Iron Age. The period of human history during which the spread of knowledge of how to make iron gradually led to the replacement of most bronze items with iron items. Since iron ore is much more plentiful than copper (the main material in bronze), this led to significant changes in civilization.

Lattice. A regular pattern that is the basis for the crystal structure of a solid. In a perfect crystal, an atom of a particular type occupies a particular site in the arrangement.

Material. A substance from which an object is made.

Materials engineering. The application of materials science to create new materials with desirable properties. See **materials science.**

Materials science. A new science that developed in the second half of the twentieth century from the parts of the older sciences of metallurgy, chemistry, and physics that relate to the microstructure of materials.

Mer. The repeating unit of a polymer molecule.

Metal. A material whose atoms are held together by metallic bonds. Metals are usually shiny, hard (when solid), and excellent conductors of heat and electricity.

Metallic bond. A bond created among all the atoms of a substance by their sharing of all the electrons in the conduction band.

Microfabrication. A collection of techniques used to make devices of microscopic size, for example, the many transistors on an integrated circuit. Many microfabrication techniques are based on precision methods of photography.

Microstructure. The arrangement of the atoms and molecules of a material, including crystal structure, defects, and differences in chemical composition.

Multiphase. Having a microstructure made up of mixed regions with different compositions, crystal structures, or both. Steel is a very important multiphase material.

Optical fiber. A long, narrow filament of glass that transmits a light beam much as an electric wire carries an electric current.

Ore. A material found in nature from which a metal can be extracted.

Photonics. A technology that is similar to electronics but uses light instead of electrons. Most experts in electronics believe that photonics will gradually expand the capabilities of electronics, and may someday replace electronics almost completely. Advances in photonics will require advances in many areas of materials science.

Piezoelectric. Having the property of producing a pulse of electric current when stressed or of responding to a pulse of electric current by changing shape. Piezoelectric materials can be used as pressure sensors or to produce high-frequency mechanical vibrations from electric currents.

Polymers. A class of materials with very large molecules made from the manifold repetition of a basic chemical unit called a mer. Plastics, synthetic fabrics, nonstick coatings, and superadhesives are among the many useful polymeric materials.

Quantum mechanics, or **quantum theory.** A field of physics that deals with the behavior of the most fundamental units of matter and energy. It has been used to understand phenomena in electronics and superconductivity, and is applied in devices such as the scanning tunneling microscope, which produces images of individual atoms and molecules.

Quenching. The technique of rapidly cooling a material to produce a desired microstructure.

Refractory. Able to withstand extremely high temperatures. One major application of ceramics is as refractory materials.

Semiconductor. A material which has an electrical conductivity that is very sensitive to the presence of impurities. Careful doping of semiconductors is the basis of the electronics industry and promises to be the basis of the photonics industry. See **doping.**

Silicate. A material containing the silicate ion (a silicon atom, three oxygen atoms, and two electrons). Many important ceramics are silicates.

Smelting. A process to remove a metal from its ore by heating in air, sometimes requiring the addition of another substance.

Stone Age. A very early period of human development that began about two million years ago and continued until about seven thousand years before the present. During most of the Stone Age, people made tools from natural materials, such as stone. In the last few thousand years of the Stone Age, they discovered how to make ceramics, the first man-made materials, from clays.

Superconductor. A material that conducts electricity perfectly without loss of energy. To date, superconductivity has been observed only at temperatures far below room temperature.

Thin film. A material made in the form of a thin coating of a substance on top of another substance. A thin film may have unusual and valuable properties because the molecules at the surface of a material behave differently than those in the interior. Thin-film materials are an important and growing area of technology.

Transistor. An electronic device that can behave as a controllable switch or an amplifier. It is made by careful doping of a semiconductor. The invention of the transistor is generally considered to have led to great technological and societal changes.

Vacancy. A crystal defect in which a lattice site is empty rather than containing the kind of atom that it would if the crystal were perfect.

X-ray crystallography. A technique in which a beam of X rays reflects from a crystal, forming a photographic image that reveals the basic crystal structure and the spacing between the atoms in the lattice.

INDEX